D1730013

A Real-time Approach to Distillation Process Control

A Real-time Approach to Distillation Process Control

Brent R. Young
University of Auckland
New Zealand

Michael A. Taube
S&D Consulting, Inc
Houston, Texas, USA

Isuru A. Udugama
The University of Waikato
New Zealand

This edition first published 2023.

Registered Office
John Wiley & Sons, Inc., 111 River Street, Hoboken, NJ 07030, USA

For details of our global editorial offices, customer services, and more information about Wiley products visit us at www.wiley.com.

Wiley also publishes its books in a variety of electronic formats and by print-on-demand. Some content that appears in standard print versions of this book may not be available in other formats.

Library of Congress Cataloging-in-Publication Data
Names: Young, Brent R., author. | Taube, Michael A., author. | Udugama, Isuru A., author.
Title: A real-time approach to distillation process control / Brent R. Young, University of Auckland New Zealand, Michael A. Taube, S&D Consulting, Inc, Isuru A. Udugama, The University of Waikato, New Zealand.
Description: Hoboken, NJ, USA : Wiley, [2023] | Includes bibliographical references and index.
Identifiers: LCCN 2022051966 (print) | LCCN 2022051967 (ebook) | ISBN 9781119669210 (cloth) | ISBN 9781119669241 (adobe pdf) | ISBN 9781119669272 (epub)
Subjects: LCSH: Distillation. | Chemical process control.
Classification: LCC TP156.D5 Y58 2023 (print) | LCC TP156.D5 (ebook) | DDC 663/.506–dc23/eng/20230111
LC record available at https://lccn.loc.gov/2022051966
LC ebook record available at https://lccn.loc.gov/2022051967

Cover Image: Wiley
Cover Design by © 06photo/Shutterstock

Set in 11.5/13.5pt STIXTwoText by Straive, Pondicherry, India

Dedicated to those from whom we (and you, the reader) have benefited:

William Y. Svrcek, PhD
Robert V. Bartman, PhD
Robert D. Kirkpatrick, PhD

Contents

Preface

There is a gap between academic study and practice of the subject of distillation process control: Students are currently unable to apply the theory as it is taught in the traditional way to the real world as they find it.

Despite the development of digital simulation tools, the subject of control theory has largely continued to be taught as it was in the 1960s or even earlier using transfer functions, frequency-domain analysis, and Laplace transforms. For linear single-loop systems such as electromechanical devices, this approach is well suited. As an approach to the control of distillation processes, which are characterized by nonlinearity, having multivariable influences, and exhibiting "slow" dynamic behavior (e.g. lag time and dead time), classical control techniques have significant limitations.

In today's rich digital environment, hardware and software are generally available in universities and workplaces to implement a "hands-on" approach to distillation process controls analysis, design, and operations using process simulators. Students and engineers are now able to experiment with digital twins or virtual plants developed within these process simulators that capture the important nonlinearities, dynamics, and multivariable nature of the real distillation units and are able to test even the most complex of control structures without struggling with oversimplified and nonintuitive mathematics or placing real plants at risk.

Lastly, while synthesis reactions are an essential element of every process plant and provide the feedstock to distillation units, due to the vast array and the specific characteristics of the many types of chemical synthesis processes (including biochemical, oil and gas processing, petrochemical, and specialty chemical), including "control" of the reaction elements is beyond the scope of this book. The common unit operation in all these processes is of course, distillation. Furthermore, distillation comprises the most flexible and manipulative handle to affect the economics of all profess units. Hence, the "reason why" for this book.

The basis of this text is therefore to provide a practical, hands-on introduction to the topic of distillation process control by using only time-based representations of the process and the associated instrumentation and controls. This book adopts the approach pioneered in *A Real-time Approach to Control* (Svrcek et al. 2000), which was the first to treat the topic of process control generally without relying at all upon classical, frequency-domain techniques such as Laplace transforms. In summary, how our book stands out is as follows:

- Comprehensive treatment of distillation fundamentals and basic to modern advanced controls from a practical point of view.
- A systematic build up from known base controls to more modern control practices with reasons explained and practical insights given.
- A break from the traditional Laplace transform approach to a time-based approach with a good mix of fundamentals and practical insights.
- A teaching text with up-to-date process simulation exercises.
- Employs an active learning, hands-on, real-time approach to facilitate student learning via process simulation workshops.
- Process simulation exercises are designed to be simulator agnostic so that they can be performed on the process simulator locally available.
- Combines control structure design and controller tuning in one handy book enabled through the unique "real-time approach."
- The hands-on exercises allow a student who follows the book to design and tune controls on a distillation column.

This textbook is designed as an introductory course on distillation process control for senior undergraduate or graduate university students in the chemical engineering curriculum. It is also a useful supplementary text to a general process control course, a primary text for a semester-long, specific distillation process control course, and as a supplementary text for a plant design course.

It is also intended that professionals, including engineers, industrial scientists, and technicians, will benefit immensely from the book, especially but not only those new to the important field of distillation control. It is expected to also form the basis of an excellent three- or four-day short course.

We believe the era of real-time, simulation-based instruction of distillation process control has finally arrived. We wish you every success as you begin to learn more about this mature and yet ever-changing and exciting field. Your comments on and suggestions for improving this textbook and workshops are solicited and are most welcome!

Reference

Svrcek, W.Y., Mahoney, D.P., and Young, B.R. (2000). *A Real-time Approach to Process Control*, 1e. Wiley.

About the Companion Website

This book is accompanied by a companion website.

www.wiley.com/go/Young/DistillationProcessControl 

This website includes:

- Selected Answers to Tutorial and Self-Study Questions for Instructors
- Figures from the book in PowerPoint

1 Introduction

1.1 The Purpose of Process Control

In both academic and industrial treatises on process control, the stated purpose for the design of control schemes is to reduce process variation, as illustrated in Figure 1.1.

While this is a true statement, it falls well short of giving a full and complete purpose that is worthy of study and mastery. In addition to the purely mathematical treatment given by most academic (as well as industrial) control courses, the would-be process control engineer has little, if any, guidance on what specific outcomes should result from proper controls design. Thus, we submit the following definition for the purpose of process control:

> *To stably, robustly, and predictably maintain Product Qualities in the face of Measured AND Unmeasured Disturbances with the* ***LEAST*** *total and incremental energy input (i.e. minimal movement) to the Process.*

Based on this definition, one begins to understand that there are several elements or attributes that must be considered regarding the proper design and implementation of control schemes. Figure 1.2 illustrates the four bodies of knowledge, which, in the authors' experience, are required for one to be a fully qualified and competent process control engineer.

Entire tomes can be (and have been) written on each of these topical areas. This book, however, while not providing a comprehensive coverage, addresses the first two and portions of the third elements, defined

A Real-time Approach to Distillation Process Control, First edition. Brent R. Young, Michael A. Taube, and Isuru A. Udugama.

Companion website: www.wiley.com/go/Young/DistillationProcessControl

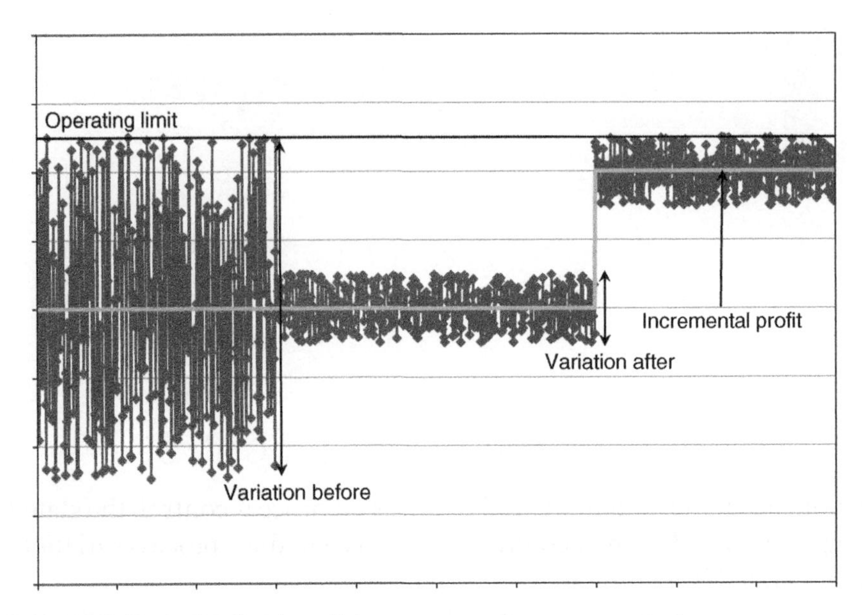

Figure 1.1 Typical definition of the purpose of process control.

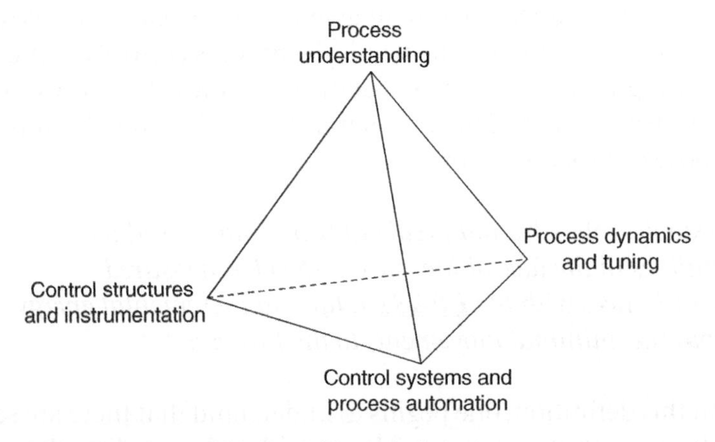

Figure 1.2 Process control bodies of knowledge.

below, to give the engineering student and practicing engineer sufficient insights such that the real-time performance of the actual distillation plant achieves (and even excels) the intended objectives. The working definitions for each of these bodies of knowledge are as follows:

- Process Understanding: By far, the singularly most important area, it encompasses all of the fundamental aspects of fluid dynamics, heat and material transfer, thermodynamics, and reaction kinetics, that is basic process engineering design and operations principles.

- Control Structures and Instrumentation: The former element being the major focus of this book. The latter is also a vital aspect, as the choice of measurement and final control devices play a significant role in the real-time behavior of all the control schemes and these topics are introduced in the text.
- Process Dynamics and Tuning: This body of knowledge is what separates the professional from the novice, in that, while the process engineer's focus is on the steady-state aspects of a process unit's design and operations, only by knowing how a process behaves in real time, that is how measured and unmeasured disturbances propagate through process over time, and then understanding how this dynamic behavior affects regulatory controls such as PID controls (as well as model predictive) controls, the process engineer can ensure that the plant achieves the intended objectives. To be clear, the "black art" of proportional-integral-derivative (PID) and model predictive control (MPC) tuning is *not* addressed herein: due to the variations, nuances and inconsistencies of how control algorithms are implemented across the many control system platforms; that is a topic requiring its own treatment. Detailed tuning is also often not the domain of the process engineer. However, pointing out the potential effects of disturbances and how they should be addressed is covered in this text.
- Control Systems and Process Automation: The proliferation of computer-based control systems has resulted in a wide array of software products and systems' capabilities and features for addressing important aspects of real-time process operations, including a dizzying array of function blocks, high-performance human–machine interfaces (HMI) (i.e. control system operating graphic displays), alarm rationalization and management and control systems design, implementation and maintenance, in addition to cybersecurity measures. Each of these aspects entail significant effort and training. While outside the scope of this book they are, nonetheless, important and vital considerations for robust, resilient, reliable, and secure process operations, and the general aspects of these topics are introduced to the reader.

While the underlying details of each of the preceding four bodies of knowledge can be expanded upon as technology improvements and enhancements are developed, the authors provide the following *Eight Rules* for successful controls implementation to assist the student and practicing engineer with guidance on what elements require focused attention:

1. Know how the process is operated and how it behaves in real time. While seemingly self-explanatory, this rule implies that one must have an intimate understanding of how a specific process unit behaves across its entire operating envelope, including startup and shutdown conditions and knowing equipment design effects (and their limitations) in various operating scenarios (i.e. maximum throughput versus turndown, product selection/optimization, and seasonal influences).
2. Understand all of the primary, secondary, and tertiary operating objectives. Again, seemingly self-explanatory but is necessary to ensure that the controls' design supports (and does not conflict with) these objectives to ensure real-time achievement.
3. Know the control system's capabilities (and limitations) and use it to its fullest; avoid "rolling your own" algorithms(s), unless absolutely necessary.
4. Document the intent of the controls and the reason(s) *why* particular functions or features are utilized. This is both for your own reference but, more importantly, for the person who takes over from you in the on-going maintenance and development of the controls you design.
5. Make the controls design reliable, resilient, robust, and maintainable. The effort required to accomplish this exceeds the typical "safe and operable" criteria applied by many engineering and operations management organizations. But it is required to ensure both the immediate and long-term achievement of the unit's objectives. The essence of this rule is that the controls should behave the same (even predictably) regardless of the conditions (throughput, weather, feed composition, etc.) to which it is subjected.
6. Make a good design into a great implementation with tuning. This is a significant element that separates the professional from the novice and, as described above, detailed tuning is substantially outside the scope of this book. Nevertheless, this is, more

often than not, the reason for poor process performance, in spite of good controls structural design.

7. Provide an intuitive interface for the operator. If/when, in the event the operators must intervene, they should be provided with a readily available (e.g. intuitive) means to "take control" of the process without having to "drill down" through control system detail displays to alter some specific parameter(s) in one, or more, of the control scheme's function blocks.
8. Identify possible or necessary follow-up enhancements for the controls. This includes addressing nonlinear behavior (a significant issue that is detrimental to all controls – both PID and model-predictive technologies) – as well as equipment changes, such as instrumentation selection and process piping and mechanical alterations.

These rules are summarized in what the authors propose should be the process control engineer's *Prime Directive*:

Prevent the controls from doing anything unexpected!

1.2 Introduction to Distillation

Distillation is the process of the physical separation of components in a liquid mixture by heating and then cooling. This is accomplished by utilizing the differences of relative volatility between the mixture components. Distillation may be used to almost completely separate components into nearly pure component products or to partially separate components such that it selectively concentrates specific components into the products.

Distillation is a unit operation of significant industrial importance. For example, in 2019, there were 132 operating refineries in the United States with a crude distillation of 18.7 million US barrels per day (US Energy Information Administration 2019). The energy use of industrial distillation also represents a significant fraction of energy usage in the chemical and process industries. White (2012) reports distillation amounts to 40% of the total energy used to operate plants in the refining and bulk chemical industries. Thus, improving the control and energy efficiency of this unit operation is important to achieving overall energy savings.

Distillation has many applications. Throughout the Hydrocarbon Process Industry (HPI), distillation is used both in upstream and downstream processing. Crude oil is stabilized by partial distillation for safe storage and transport, and, at the refinery, fractional distillation is used to separate it into fuel products and chemical feedstocks. In addition to refining, distillation is used industrially in many other applications. Distillation is used in the chemical industry to separate and purify chemical reaction products to produce sales streams. The distillation of the products of fermentation and other bio-industry processes also produces many products of commercial value, including distilled beverages of high alcohol content. Distillation is also used in cryogenic processes to separate air into its constituent components for use in industrial and medical grade gases, as well as for liquified natural gas (LNG) obtained from the associated gas out of oil and gas wells. And distillation continues to be used as a desalination treatment solution. It is a small wonder that it has been estimated that there are more than 40 000 distillation columns in North America (White 2012).

Distillation has a long history across most, if not all, ancient cultures. Early evidence of distillation has been found on Mesopotamian Babylonian tablets from around 1200 BCE that described distillation for perfumery operations (Levey 1959). Distillation may have been practiced in China as early as the first century of the Common Era (CE) (Haw 2012). Evidence has also been found in Roman (Forbes 1970) and Byzantine (Bunch and Hellemans 2004) Egypt in the first and third centuries CE. Distilled water has been produced since at least 200 CE (Taylor 1945). Distillation was also used to make weak liquor in ancient India (modern Pakistan) in the early centuries of the CE (e.g. Husain 1993).

Medieval Arabic chemists worked on the distillation of various substances from the eighth century of the CE (al-Hassan 2009). By the twelfth century, fractional distillation (Burnett 2001) and the production of ethanol by the distillation of wine with salt (Multhauf 1966) had become known to Western European chemists.

As human history moved from the agricultural era to the first industrial revolution in the nineteenth century, the basis of modern distillation processing was developed – continuous processing, reflux, trayed columns, and preheating (Othmer 1982).

Chemical engineering's genesis as a separate discipline at the end of the nineteenth century provided a scientific foundation to the development of distillation, simultaneously with the second industrial revolution and the developing petroleum industry with the development of design methods, including the Fenske equation (1932), the McCabe–Thiele method (1925), and the unit operations approach (Hougen 1977) in general.

Subsequent to these developments, the advent of heat integration, pinch technology, and process intensification for energy efficiency since the 1970s resulted in more complex and alternative distillation column design proposals, which, in some cases, have been implemented in industry, such as divided wall, reactive distillation, and Petlyuk columns (e.g. Doherty and Malone 2001).

A few words about how calculations are performed before talking about distillation process control are appropriate for a book advocating a real-time (or simulation-based) approach. Before the 1950s, calculations were done manually (e.g. using a slide rule with pencil and paper). As computer technology became more accessible, these manual calculations were implemented using simple programming languages, such as FORTRAN and, later, BASIC as the personal computer (PC) revolution came into being during the third industrial revolution in the 1970s and 1980s. Finally, today's fourth industrial revolution (Industry 4.0) has ushered in a plethora of process simulation software for the fast, accurate steady-state, as well as dynamic, simulation (e.g. as described in detail in Svrcek et al. 2014), and development of "digital twins" of distillation columns and entire process facilities – such tools as Aspen HYSYS, Schlumberger's Symmetry, and Siemens PSE's gPROMS. So, while manual methods still have their place for initial conceptualization and sanity checks, digital modeling tools enable fast, accurate, and precise modeling of complex processes and unit operations.

1.3 Distillation Process Control

The traditional approach to general control loop analysis and design for all processes, including distillation, was based on mechanical and electrical engineering methods very often derived from the frequency domain, such as transfer functions, Laplace transforms, Bode plots, and Nyquist diagrams. While perhaps somewhat helpful for developing a deep understanding of dynamics, these methods are essentially abstract mathematical formulations and were primarily pen and paper techniques for solving linear ordinary differential equations for single-loop systems when the computational resources were unavailable.

By knowing or identifying the Laplace transfer function of the process, a student or engineer could then use a range of controller tuning resources to create an ideal controller equation. However, this approach had major drawbacks that included:

- The mathematical concept was often too abstract for students to grasp and apply practically.
- An apparent disjoint between Laplace with the "time-domain" often results in a lack of intuition for users.
- They use linearized systems equations that oversimplify the complexities of distillation dynamics.
- They are difficult to apply to real, multiple input and output processes, such as when attempting to develop multiloop controls.

Educationalists and industrialists alike have realized these issues and limitations and subsequently have taken a different approach. Present-day distillation control texts tend to either be aimed at being references for practitioners with minimal or no treatment of process calculations or simulations, or be comprehensive academic texts that are similarly lacking in calculations and simulations, and sometimes only treat the fundamentals very lightly.

Notable examples of the former industry-focused texts include (i) Shinskey (1977), which was a classic in the field for its time but was not a teaching text and included no advanced control or simulation; (ii) Luyben (1992), which is an edited volume of leading industry practitioners and academics and very good for its time but not a teaching text with no examples, exercises, or simulations; (iii) Luyben (2013), which includes design and control but whose primary focus is how to use a specific simulator rather than distillation and simulation fundamentals per se; (iv) Nag (2015), which is a practical reference intended for processing engineers and not a teaching text with minimal calculations and simulations.

Academic/teaching texts addressing distillation process control include (i) Robbins (2011), which is geared toward control structure using a lesson style but has very light coverage with little dynamics considered; (ii) Smith (2012) is a comprehensive distillation control book covering lots of basic and advanced control techniques, but no time-varying behavior (dynamics), nor mention of simulation or exercises; (iii) Kiss (2013) that focused on advanced distillation arrangements and is comprehensive in that regard, but has cursory coverage of fundamentals, basic distillation, controls, and advanced process control; (iv) Gorak and Schoenmaker (2014) who present a comprehensive, well-written coverage of the topic, but without sections on advanced process control, light coverage of simulation, and no explicit coverage of energy usage.

1.4 A Real-Time Approach to Distillation Process Control Education

With the ever-improving computational capabilities, high-fidelity distillation modeling (and control) emerged in the late 1980s. These tools explicitly modeled the mass transfer, transport phenomena, thermodynamics, and the vapor–liquid equilibrium (VLE) of a given distillation column. By the early 2000s industrial process simulators were able to accurately and explicitly model the time-dependent behavior of a distillation column (i.e. dynamics). The simulation environment is graphical in nature, somewhat mimicking a distributed control system's (DCS) graphical screen, enabling even modestly experienced users to set up a distillation process with relative ease. The implications for education are that the users could focus on learning the development of controls rather than how to represent a distillation operation using abstract mathematics such as Laplace Transforms; refer to the book *A Real-Time Approach to Process Control* (Svrcek et al. 2014) that includes a step-by-step guide on how this was achieved. The main benefits of applying a simulation-based approach to teaching and learning distillation control for the user/students are as follows:

- Users/students experience realistic nonlinear, coupled process responses that are expected from a distillation column, as opposed to idealized linear, decoupled mathematical functions, as there is a one-to-one correspondence between the dynamic distillation models and the actual plant.
- Users/students are able to learn how to set up sensors, controllers, and valves to achieve both an overall as well as specific control objective(s).
- Through interaction with the dynamic process simulator, users/students can develop an intuitive understanding of the dynamics

of distillation control and relate learnings from other areas of process engineering to the observations.

- Users/students are able to compare candidate control structures and assess the propagation of disturbances through a distillation plant, enabling the evaluation of advanced process control and plant-wide control schemes.

Before summarizing the content of the following chapters, it is appropriate to mention one thing that this book consciously does not cover in detail – instrumentation and control hardware. The hardware of course makes software happen and is therefore very important for control scheme execution. In Svrcek et al. (2014), a broad introduction is given. However, this topic is worth a book in itself, and indeed, there is an excellent text that provides a very comprehensive cover (McMillan and Vegas 2019).

This text is organized into a framework that provides relevant theory and industrial experience, along with a series of hands-on workshops that employ computer simulations that test and explore the theory. This chapter provided conceptual practical basis and a historical overview of the field of real-time distillation process control, including simulation, and the pros (mainly) and cons of a real-time approach. Chapter 2 introduces the fundamentals of distillation control, covering the basic principles of distillation and relevant control information, including pressure and inventory control. Chapter 3 introduces the reader to the fundamentals of hardware that the "software" described in this book runs on. In Chapter 4, we look at distillation inventory control, a key set of issues for stability, in greater detail. Then, in Chapter 5, we look at composition control. Gain analysis, inferential and analyzer control, common control structures, batch distillation, and energy controls are introduced here. Chapter 6 contrasts historical refinery versus chemical plant distillation control and the applicability of traditional refinery and chemical distillation control structures. Chapter 7 focuses specifically on distillation control tuning. Topics covered include integrating versus open-loop stable processes, when to apply tight versus loose control, and handling equipment constraints via tuning. Fine chemical distillation control is examined in Chapter 8 that discusses the key features of nonlinearities, measurements, and basic side draw column controls. Chapter 9 looks at advanced regulatory control – namely feedforward, cascade, and ratio control – and advanced side draw column controls.

Chapter 10 tackles more complex control methods: multivariable model predictive control of distillation columns. Finally, in Chapter 11, we take a look at some of the important issues related to plant-wide control problems, including distillation. A combination of theory and applied methodology provides a practical treatment to this complex topic. And the chapters are matched by a series of simulation workshop exercises that serve to illustrate, teach, and test the concepts developed in each chapter.

Tutorial and Self Study Questions

1.1 What is the fundamental purpose of process control (that does not use "control" in the description)?

1.2 What questions are most important for a process control engineer (PCE) to address when assessing a (new) process and its control design (compared to typical management questions)? What is the effect on time and effort?

1.3 What are the three most important items required for any (complex) control scheme and when/how should it be generated?

1.4 What three (or four) bodies of knowledge are required for a person to be fully competent in process control and what is the order of importance (1 = Most Important)?

References

Bunch, B.H. and Hellemans, A. (2004). *The History of Science and Technology*, 88. Houghton Mifflin Harcourt.

Burnett, C. (2001). The coherence of the Arabic–Latin translation program in Toledo in the twelfth century. *Science in Context* 14 (1–2): 249–288.

Doherty, M.F. and Malone, M.F. (2001). *Conceptual Design of Distillation Systems*. McGraw-Hill.

Fenske, M.R. (1932). Fractionation of straight-run Pennsylvania gasoline. *Industrial and Engineering Chemistry* 24: 482.

Forbes, R.J. (1970). *A Short History of the Art of Distillation: From the Beginnings Up to the Death of Cellier Blumenthal*, 2e. Leiden: Brill Publishers.

Gorak, A. and Schoenmaker, H. (2014). *Distillation: Operation and Application*. Academic Press.

al-Hassan, A.Y. (2009). *"Alcohol and the Distillation of Wine in Arabic Sources from the 8th Century". Studies in al-Kimya': Critical Issues in Latin and Arabic Alchemy and Chemistry*, 283–298. Hildesheim: Georg Olms Verlag.

Haw, S.G. (2012). "Wine, women and poison".). *Marco Polo in China*, 147–148. Routledge.

Hougen, O.A. (1977). Seven decades of chemical engineering. *Chemical Engineering Progress* 73: 89–104.

Husain, J. (1993). The so-called 'Distillery' at Shaikhan Dheri – a case study. *Journal of the Pakistan Historical Society* 41 (3): 289–314.

Kiss, A. (2013). *Advanced Distillation Technologies: Design, Control and Applications*. Wiley.

Levey, M. (1959). *Chemistry and Chemical Technology in Ancient Mesopotamia*, 36. Elsevier.

Luyben, W.L. (ed.) (1992). *Practical Distillation Control*. Van Nostrand Rheinhold.

Luyben, W.L. (2013). *Distillation Design and Control Using Aspen Simulation*, 2e. Wiley.

McCabe, W.L. and Thiele, E.W. (1925). Graphical design of fractionating columns. *Industrial and Engineering Chemistry* 17: 605–611.

McMillan, G.K. and Vegas, P.H. (2019). *Process Industrial Instruments and Control Handbook*, 6e. McGraw-Hill.

Multhauf, R.P. (1966). *The Origins of Chemistry*, 204–206. London: Oldbourne. ISBN 9782881245947.

Nag, A. (2015). *Distillation & Hydrocarbon Processing Practices*. Penn Well.

Othmer, D.F. (1982). Distillation – some steps in its development. In: *A Century of Chemical Engineering* (ed. W.F. Furter). Boca Raton, FL, USA.

Robbins, L. (2011). *Distillation Control, Optimisation, and Tuning. Fundamentals and Strategies*. CRC Press.

Shinskey, F.G. (1977). *Distillation Control*. McGraw Hill.

Smith, C. (2012). *Distillation Control: An Engineering Perspective*. Wiley.

Svrcek, W.Y., Mahoney, D.P., and Young, B.R. (2014). *A Real-time Approach to Process Control*, 3e. Wiley.

Taylor, F. (1945). The evolution of the still. *Annals of Science* 5 (3): 185.

US Energy Information Administration (2019). Number and capacity of petroleum refineries. `https://www.eia.gov/dnav/pet/PET_PNP_CAP1_DCU_NUS_A.htm` (accessed 9 November 2019).

White, D.C. (2012). *Optimise Energy Use in Distillation*, 35–41. CEP, March 2012.

2 Fundamentals of Distillation Control

Understanding the fundamentals of distillation is the key to developing proper process control structures: the combination of fundamental process understanding, including steady-state and dynamic behaviors, and process control concepts are used to develop solutions that robustly and efficiently regulates a given process. From a fundamental point of view, a distillation column employs the differences of boiling points between two or more compounds to facilitate separation. Typically this means the heavier compounds with higher boiling points go toward the bottom of the column, while the lighter compounds, with a lower boiling point, propogate toward the top. To achieve a significant separation (or to meet the required product purity specifications), distillation columns employ multiple theoretical separation stages (as per Figure 2.1) stacked on top of each other.

In each theoretical stage, vapor coming from the lower part of the column (consisting of a larger proportion of heavier compounds) comes into contact with the liquid coming from a higher part of the column (consisting of a larger proportion of light compounds). These two streams mix, eventually reaching a thermodynamic vapor–liquid equilibrium (VLE). As a result, the vapor leaving the stage is slightly lighter than the vapor arriving. Similarly, the liquid leaving the stage is heavier than the liquid arriving onto it. Repeating this process throughout the distillation column results in the separation of the feed components.

To this end, the primary objective of distillation control structures is to generate a stable thermodynamic environment wherein the theoretical units can function at greatest efficiency and effectiveness.

A Real-time Approach to Distillation Process Control, First edition. Brent R. Young, Michael A. Taube, and Isuru A. Udugama.

Companion website: www.wiley.com/go/Young/DistillationProcessControl

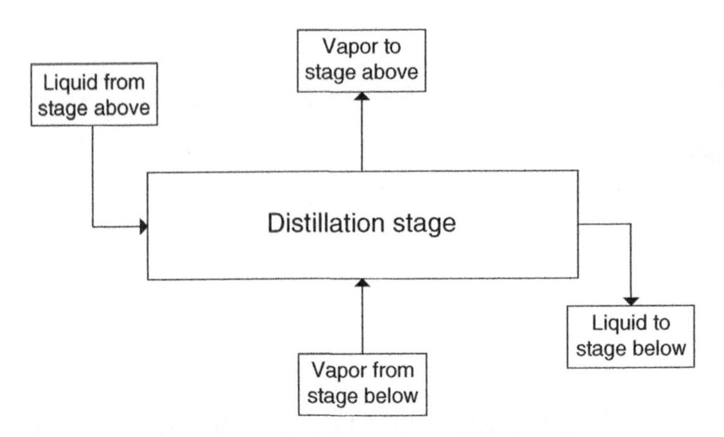

Figure 2.1 Distillation theoretical stage liquid and vapor flows.

Or, to put in other words, they exist to ensure that the vapor and liquid and variables such as column pressure (which influences the VLE) are kept at optimal targets. From a practical point of view, this requires that the column vapor and liquid inventories be managed to achieve stable operations. To be clear: a distillation column is not an "open-loop stable" process and requires controls, at a minimum, to manage fluid inventories.

Understanding the unit's operating objectives and the controls' design to achieve them should be collaborative between process engineers and control engineers. Left to either of these contributors alone, the resulting controls often fall short of the intended performance. The process engineer understands the process' design, based on the steady-state behavior, while the process control engineer leverages his understanding of dynamic behavior and proper variable pairing, as well as PID tuning, to ensure that the controls support and reinforce the desired real-time process behavior. The critical factor for successful collaboration is that the process engineer describes *WHAT* objectives are intended, while the process control engineer defines *HOW* they are achieved.

Process understanding is a key but often overlooked aspect of successful controls design. In practice, too little time is spent on analyzing process data to understand how a process really works in real time. Modeling and simulation are important elements in gaining this understanding for new or grassroots units.

While rigorous dynamic simulation provides important insights to a unit's expected real-time behavior, even simple dynamic simulations can be useful for determining a unit's response to various disturbances and how the controls should be structured to minimize the impact of the disturbances to the unit's objectives. A flexible, dynamic simulation allows for rapid evaluation of different control structures and their response to various disturbances.

2.1 Mass and Energy Balance: The Only Means to Affect Distillation Tower's Behavior

To determine the optimal control scheme for a distillation unit, several considerations must be taken into account. First, however, it is important to remember that ALL distillation columns, regardless of internals (i.e. trays, random or structured packing) or physical configuration (e.g. two-product, side-draw(s)), have two and ONLY two (2) "handles" by which they are affected: mass balance and energy balance. Also, mass and energy balance must never be confused with degrees of freedom. As the number of liquid and/or vapor draws and returns, as well as heat sources and/or sinks increase, so do the number of degrees of freedom. But there are still two, and ONLY two, *handles* that each degree of freedom potentially manipulates to affect the behavior of a distillation column.

Although the above definitions appear simple enough at face value, identifying which handle a particular control scheme is manipulating may not be quite as simple and actually depends on the design of other controls on the column – in particular the support controls (e.g. level/inventory controls). So, for additional clarity, these terms are defined as follows:

1. Mass Balance: The distribution of the incoming mass across all the available out-going streams, both on a gross and individual component basis.
2. Energy Balance: Sometimes described as "fractionation" or "internal circulation," refers to the amount of vapor AND liquid traffic within the column, which, ultimately, affects tray/packing efficiency and the ability to separate the fluid components. To further clarify, the key issue to recognize is that "Energy Balance" requires BOTH liquid and vapor traffic to be affected, not just one or the other.

In terms of the effectiveness of one handle versus the other, mass balance is the most influential. In process control terms, the process gain (*Kp*) is "large," thus requiring the least movement to affect a change. Once a minimum and stable vapor–liquid circulation (energy balance) is established on a column, the single most effective means to affect a stream's composition is by manipulating the column's mass balance.

Furthermore, as a general rule, changes in mass balance generate the fastest dynamic response for affecting a stream's composition. As a result, for any stream composition that represents a critical or "never exceed" limit, whatever controls are defined for that composition should manipulate mass balance, as this will be the fastest means by which to affect the composition. The control scheme should also be, in principle, easier to tune relative to alternate schemes that manipulate energy balance due to the faster dynamic response.

To illustrate the differences between these two handles, let's consider a hypothetical binary-split distillation tower: Once the tower is operating at a steady state, a temperature/composition profile is established, as illustrated in Figure 2.2a. The exact shape and position of the profile is determined by the feed composition, relative component VLE characteristics, and the tower's design (e.g. number of trays, tray-type and efficiency, feed and draw nozzle locations). The base regulatory controls design also depends upon the fluid physical states (saturated

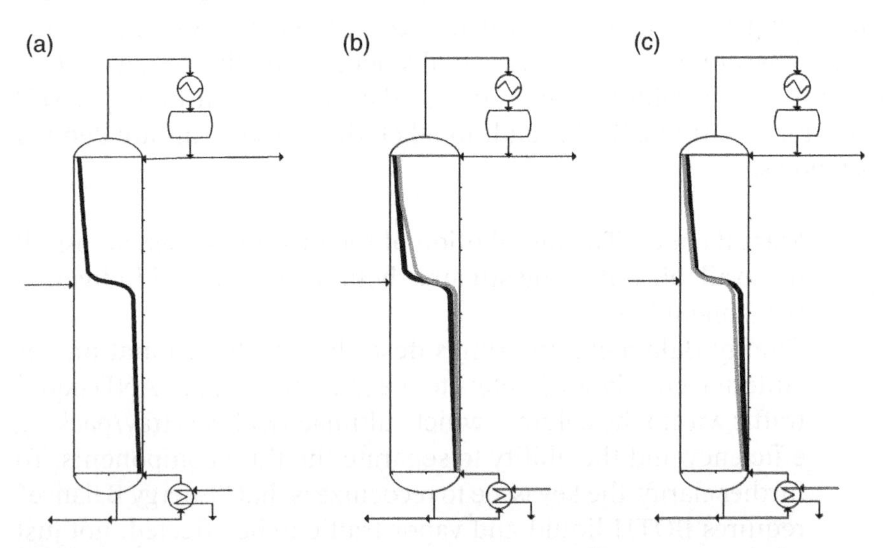

Figure 2.2 Changes to temperature/composition profile using mass balance.

or sub-cooled liquid, superheated, or saturated vapor) and available heat sources and sinks.

This hypothetical example provides a conceptual basis to illustrate how mass and energy balances affect the temperature/composition profile in all distillation columns. If the mass balance is changed by increasing the overhead distillate flow, then the profile gets shifted upward in the column, as illustrated in Figure 2.2b. Conversely, if the overhead distillate flow is reduced, then the profile shifts downward; see Figure 2.2c. While no regulatory controls are shown, it is assumed that all other process parameters (i.e. feed rate, reboiler duty) remain constant, and there is no net accumulation of mass and/or energy resulting from the described process change.

Manipulating the energy balance, however, does something different: it alters the shape of the temperature/composition profile. So, for example, if fractionation is increased from that illustrated in Figure 2.3a, then the profile "flattens" or "compresses," reflecting the effects of increased vapor–liquid contact and a relative improvement in tray efficiency, as illustrated in Figure 2.3b. If fractionation is decreased from that in Figure 2.3a, then the profile "stretches," as shown in Figure 2.3c.

There are, of course, limits to how far both mass and energy balance can affect a column's behavior, primarily due to VLE characteristics and, to a lesser extent, the column's mechanical design. And, for the

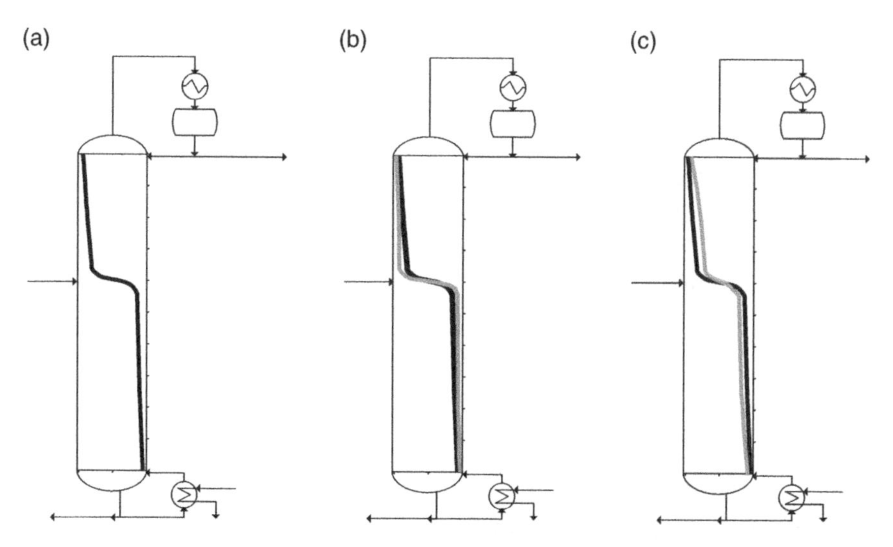

Figure 2.3 Changes to temperature/composition profile using energy balance.

sake of accuracy, it is recognized that adjustments to any MV affect both mass and energy balance to some extent. Nevertheless, their respective effects are summarized as follows:

- Changes in mass balance shift the composition/temperature profile up/down;
- Changes in energy balance stretch or compress the profile's shape.

From a theoretical basis, the following equations describe how the various manipulated variables are related. And, while virtually any variable pairing can be used to achieve the desired control objectives, a careful analysis will reveal the optimal pairing that generates the most effect with the least effort or energy input to the process.

Overall material balance:

$$F = D + B \tag{2.1}$$

Component balance:

$$Fx_f = Dx_d + Bx_b \tag{2.2}$$

where:

$F =$ feed
$B =$ bottoms
$D =$ distillate
$x_i =$ concentration of a particular component in the feed, distillate or bottoms
$Q_{reb} =$ reboiler duty
$Q_{cond} =$ condenser duty

Energy balance:

$$FH_f + Q_{reb} = DH_d + BH_b + Q_{cond} \tag{2.3}$$

where:

$H_F =$ enthalpy of the feed
$H_B =$ enthalpy of the bottoms
$H_D =$ enthalpy of the distillate

Combining Eqs. (2.1) and (2.2) to eliminate B or D gives:

$$D = F(x_f - x_b)/(x_d - x_b) \tag{2.4}$$

$$B = F(x_f - x_d)/(x_b - x_d) \tag{2.5}$$

The control structures must at all times satisfy Eqs. (2.1) and (2.3). For particular values of x_d and x_b (i.e. composition specifications), Eq. (2.4) or Eq. (2.5) also have to be satisfied. D, B, Q_{reb}, and Q_{cond} can all be fixed or adjusted dynamically by control valves on the flow rate of the utility streams. While historically the reflux flow has been used as a means to affect composition, doing so directly affects the energy balance of the tower as well as masks the influence on the material balance. In short, a distillation tower's Reflux flow is NOT an independent variable; this nuance will be discussed later.

2.2 Control Design Procedure

Prior to delving deeper into the design of distillation control structures, it is important to define a few terms. Firstly, engineers often use the terms "control strategy," "control structure," and "control scheme" interchangeably. Regardless of which phrase is used, at its core, it refers to the selection and pairing of manipulated variables (MV) and controlled variables (CV) to form a complete, functional control loop.

The following sections describe a methodology for designing control schemes that includes control strategy considerations, control structure selection, and variable pairing elements. The methodology is largely empirical and based on general principles for distillation operations. The methodology for control structure' design assumes that the process configuration is fixed and that mechanical changes are not possible. This assumption holds in many grass-roots, as well as expansion, projects as process control applications designs are only touched upon briefly at the initial design phase, while the final controller design work is performed well into the detailed design phase (and usually well after equipment mechanical designs have been or are very nearly completed, which precludes making any mechanical design changes that would otherwise enhance real-time performance).

The terms "decentralized" and "centralized" control are also used in distillation operations. Decentralized strategies commonly refer to a controller configuration where a single input (a process variable [PV] or controlled variable) is used to adjust a single output (manipulated variable). However, advanced decentralized control configurations also consider concatenated or calculated variables of importance, such as Internal Reflux, Total Duty, Pressure-Temperature Compensated flow, or Pressure-Compensated Temperature (some of which will be explained further in subsequent chapters). Conversely, centralized control strategies usually refer to multi-input (process or controlled variable) and multi-output (manipulated variable) applications, where control actions are determined based on optimization algorithms instead of individual PID controls. Examples of such applications include model-predictive or matrix control technologies, which have been in use since the 1980s.

2.3 Degrees of Freedom

The economic performance of a distillation system is linked to its steady-state degrees of freedom. In other words, the economic benefits of a column's control schemes depend on how well it achieves and maintains composition, recovery, or yield and not necessarily on how well it holds integrating variables such as levels and pressures. The integrating variables must obviously be maintained, but their control performances do not directly translate into profits. However, inventory controls can be the most troublesome of all loops and can preoccupy the operators to the point where the economically important composition and recovery are neglected. This problem has been resolved by designing the level and pressure controls before dealing with the composition controls (e.g. Buckley 1964). However, one must be careful in selecting the manipulated variables for inventory control as they can significantly impact the control performance of the composition loops.

When a process engineer works with a steady-state simulation of a distillation column, a certain number of variables have to be specified in order to converge to a solution. The number of variables that need to be specified, or degrees of freedom, are determined through the description rule as stated by King (1980):

> *In order to describe a separation process uniquely, the number of independent variables which must be specified is equal to the number which can be set by construction or controlled during operation by independent, external means.*

Applying the rule to a distillation column with a total condenser and two product streams gives two (2) degrees of freedom. In this case, the column requires two specifications (i.e. a composition and a component recovery). The steady-state simulator will then manipulate two variables, such as reboiler duty and one product stream flow, in order to satisfy the specifications, while the condenser duty and other stream's flow will be determined by closing the material and energy balances.

When the same two-product distillation column is transferred from a steady-state simulation to a dynamic simulation, the degrees of freedom increase from 2 to 5. These three new degrees of freedom correspond to three inventory variables within the column that are essentially ignored by the steady-state material and energy balances: from a steady-state perspective, these variables do not exist; this concept is covered further in Chapter 4. The inventory variables for this column are the condenser (overhead accumulator) level, reboiler (tower bottoms accumulator) level, and the column pressure, and they represent restrictions on, or limits to, the control of a distillation column, as follows:

- The overall enthalpy balance limits the heat removed by the condenser and added by the reboiler.
- The rate of distillate and bottoms produced may not exceed the feed rate.
- The number of stages in the column and the reflux ratio must be greater than or equal to the number required for the desired separation (King 1980).

In order to maintain these inventory variables, as well as the mass and energy balances, a base regulatory control layer must be established. From a practical point of view, the regulatory control layer consists of decentralized controls where one control valve (or degree of freedom) is used as the manipulated variable (MV) to maintain each controlled variable (CV). This relationship between CVs and degrees of freedom (control valves or MVs) is known as variable pairing, or MV–CV pairing, and is an important concept in control structure design. This decentralized approach is best performed at the regulatory control layer as it is a simple and robust control structure and its objective of "regulating" the process does not require any optimization. Centralized controls can be later added on top of this regulatory control layer where the manipulated variable of the centralized controls are the set-point of the regulatory controls.

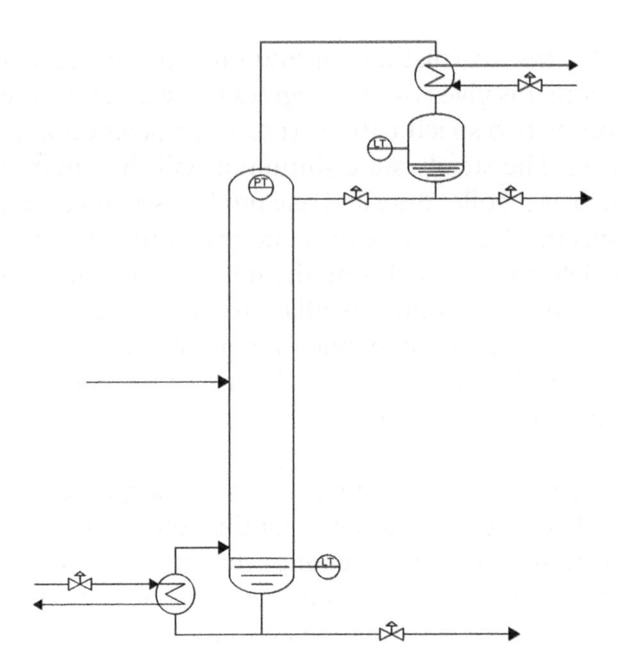

Figure 2.4 Distillation column schematic with five manipulated variables (the control valves).

When the five manipulated variables, which correspond to five valve, as shown in Figure 2.4, are viewed, it is observed that the two steady-state manipulated variables are a subset of the overall five. It is important to note, however, that the heat duties are not exclusively steady-state manipulators, which prevent them from being used for dynamic inventory control.

To clarify: while the first pass of the controls structures' design often uses Reboiler/Condenser Duty as the MV for maintaining energy balance, as the mechanical design of the equipment is completed, these basic control structures, specifically the MV–CV Pairings, must be assessed with a relative gain analysis (RGA) (covered in Section 2.5). The results of the RGA may reveal inadequacies in the original structure designs (i.e. the process gain(s) is so small that the controller must make very large movements in order to affect the CV), thus prompting a redesign (i.e. changing certain MV–CV pairings). And to be clear: this is strictly a steady-state analysis (no consideration of dynamic behavior is required) to determine which MV–CV pairing(s) will satisfy the *purpose* of process control (as given in Chapter 1): To get the greatest effect with the least energy-input (or movement) to the process! The implications for this assessment are covered in greater detail in Chapter 4.

Although the previous paragraphs describe the manipulated variables as control valves, there are many choices available other than just the individual valves or flows they represent. For example, many columns have ratio controls as a manipulated variable (e.g. Duty/Feed, Distillate/Feed). When ratios and linear combinations of variables are included, the choice of a manipulator for a given loop broadens for a simple two-product column. However, the steady state and dynamic degrees of freedom remain unchanged as two and three, respectively, totaling five.

From a practical (as well as theoretical) perspective, however, the Reboiler Duty and Reflux flow variables are not independent: they are *interdependent*. Thus, one cannot independently set both of these variables. This concept is illustrated in Figure 2.5 and explained in detail, below.

If one views distillation from the perspective of a "black box," that is not knowing what is inside the box and knowing only the inputs and outputs, then one observes a distillation column as shown in Figure 2.5a: the feed stream and energy input to the column as inputs and two (2) product streams as outputs. However, knowing that mass balance must be maintained, one realizes that the two product streams are not independent of each other and their movements must be coordinated, as illustrated in Figure 2.5b: as one stream is increased, the other must decrease and vice versa. Lastly, to give context to this concept, a typical 2-product distillation tower is placed inside the box, as shown in Figure 2.5c. Here one observes that, as discussed above, Reflux is

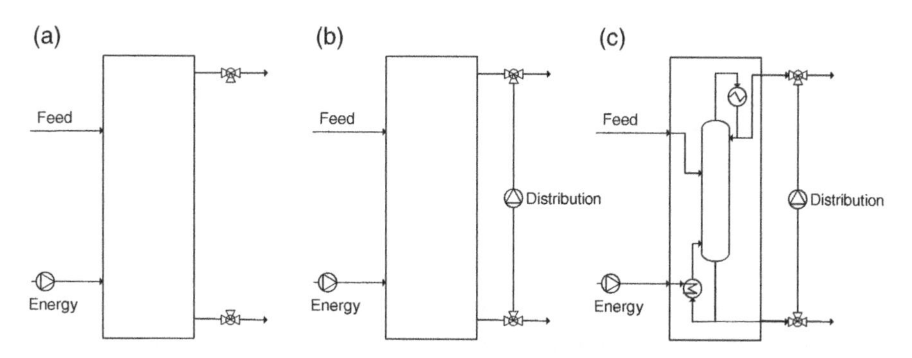

Figure 2.5 Distillation column as a "black box."

not, in fact, an independent variable as it is "inside the box" and not available for independent manipulation.

In summary, the total degrees of freedom for actual plant operations equals the number of independent flows (valves) that are available for manipulation in that section of the plant. To find out how many integrating variables (i.e. pressures and levels) are to be controlled with the available valves, subtract the degrees of freedom required for steady-state control from the total degrees of freedom. In a typical total condensing distillation column, there are the following valves or streams:

1. Tower bottoms flow
2. Reboiler duty
3. Distillate flow
4. Condenser duty
5. Reflux flow

These streams must be used to control the following integrating variables

1. Overhead Pressure
2. Condenser/Overhead Accumulator Level
3. Reboiler/Bottoms Accumulator Level

As such, in this system only two (2) streams will be left for controlling the column after subtracting the three (3) streams that must be dedicated to the control of the three (3) integrating variables.

2.4 Pairing

One of the most vital aspects of distillation control is the interaction between the material and energy balance controls and their effect(s) on stream composition. Depending on the inventory controls' structures, heat input or removal can alter both the material draws and the compositions. This interaction can work in support of or against the column's operating objectives, thus the importance of understanding the tower's operating objectives! A key requirement of any control scheme is that it relates directly to the process objectives. A control structure that does not meet the process objectives or produces results that conflict with the process objectives does not add value to the process. This topic will be covered further in Chapter 5.

Another point to consider when choosing a column control scheme is that, typically, the process gains in high-purity separations are very nonlinear. This is verified by simply using the component material balance equations. For example, Eq. (2.4) can be rearranged and differentiated at constant x_B to give:

$$\left(\partial x_D / \partial D\right)_{x_B} = K_{x_D} = -\frac{F\left(x_f - x_B\right)}{D^2} \tag{2.6}$$

Equation (2.6) shows that changes in the distillate rate, D, will have a much larger effect on the distillate composition, x_d, when the distillate rate is relatively low as compared with cases when the distillate rate is relatively high. This nonlinear relationship is illustrated in Figures 2.6 and 2.7.

Figure 2.6 shows the relationship between distillate composition, x_D, versus the distillate flow, D (as a fraction of the feed flow), for various values of feed composition, x_f. A second-order polynomial curve fit is also shown next to each curve, representing fractional feed compositions of 0.5, 0.25, 0.1, and 0.01. In all cases, the curves cover the entire feasible range of distillate flow (i.e. 0–100% of feed rate). Figure 2.7

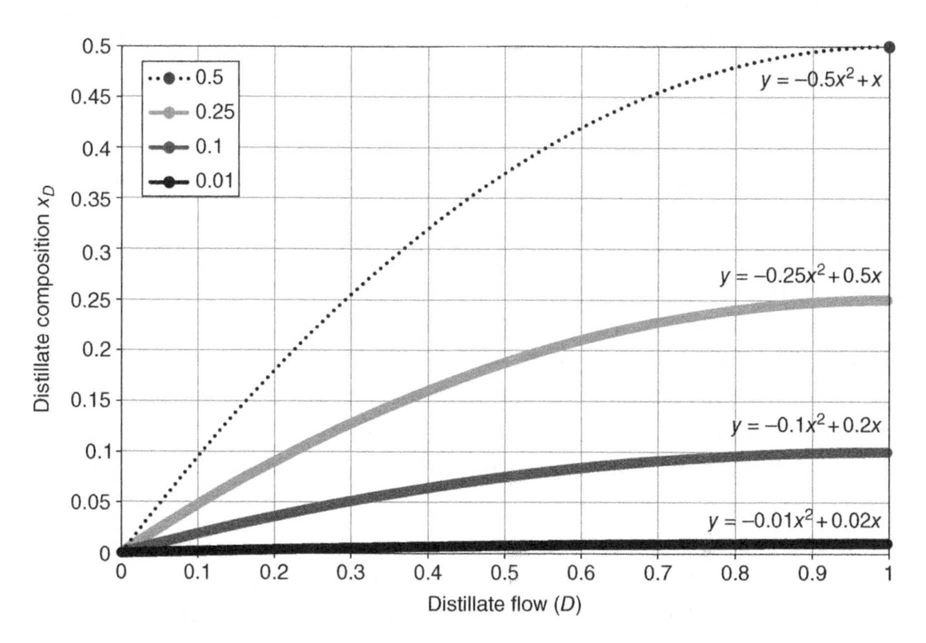

Figure 2.6 Plot of distillate composition versus distillate flow.

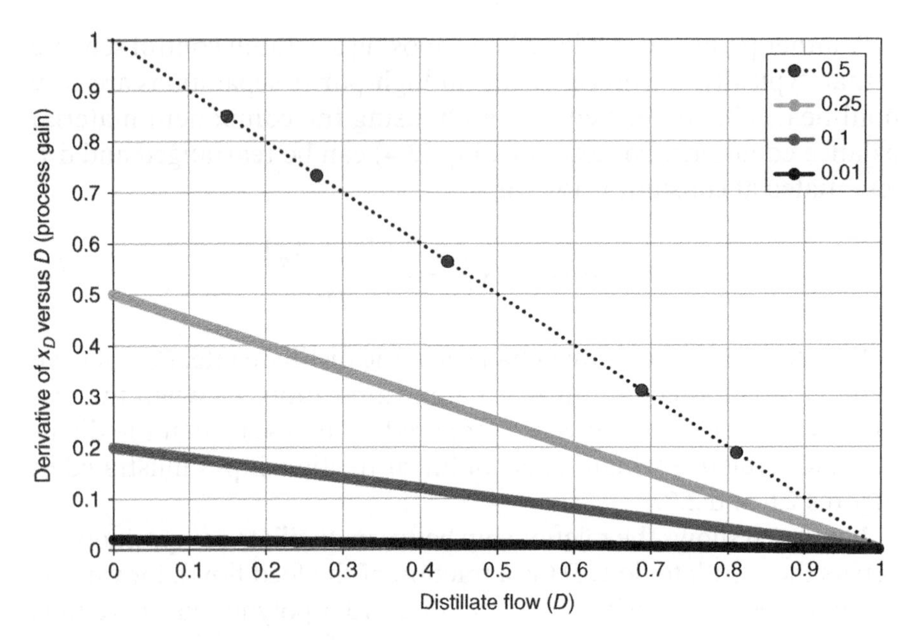

Figure 2.7 Plot of distillate composition versus distillate flow.

shows the *derivative*, with respect to distillate, for each of the curves in Figure 2.6, which represents the change in *process gain* (K_p) as a function of the fractional distillate flow. As seen in Figure 2.7, the largest effect of changes in distillate flow on the distillate composition is at (or near) zero; the larger distillate flow becomes, the less influence it has on distillate composition.

A final and important consideration to keep in mind is the dead time that may manifest for certain stream compositions. It is easy to see how a distillation column with its multiple stages can exhibit dead time behavior depending on which independent variable (flow, valve, or ratio) is changed and the dependent variable change is observed. Hence, it is vital that the control structures on a distillation column be set up to minimize the dead time response vis-à-vis the measured and unmeasured disturbances.

The steps for determining a suitable control structure are as follows:

1. Define the operating objectives of the process (e.g. stream quality, utility usage, throughput) and the nature (magnitude, duration, frequency, etc.) of the measured and unmeasured disturbances.

2. Understand the process behavior in terms of both its steady-state and dynamic behavior.
3. Select the MV–CV pairings that directly support and/or achieve the operating objectives, keeping the following criteria in mind:
   a. Select an MV that is physically closest to the CV in the process; and
   b. Has the largest steady-state *process gain* (and, albeit a secondary objective, the fastest dynamic behavior) compared to other potential MVs.
4. Configure the controls and evaluate them with anticipated disturbances through the use of dynamic simulation.

Ultimately, the importance of process control is seen through increased overall process efficiency, which allows the plant engineer and operator to get the most from the process' design. This is especially true of distillation. Due to overdesign, many older distillation columns are very flexible in terms of product yields and compositions, as well as varying levels of energy input. Thus, one finds many such columns operating at close to 200% of their original design throughputs!

The following methodology (Tyreus 1992) is recommended to define a control structure for a simple distillation column shown in Figure 2.4.

1. Count the control valves in the process to determine the overall degrees of freedom for control.
2. Determine the steady-state degrees of freedom from a steady-state analysis.
3. Subtract the steady-state degrees of freedom from the overall degrees of freedom to determine how many inventory loops can be closed with available control valves.
4. Design pressure and level controls and then test for disturbance rejection.
5. Design composition controls based on the product stream requirements. It is important that the manipulated variable chosen can control the feed split.
6. Design optimizing controls with the remaining manipulative variables, if any.

There are five degrees of freedom for the simple distillation column in Figure 2.4, which translates into five independent valves from a control point of view. In this 5×5 system, there are 120 possible single input/single output (SISO) control combinations of controlled and manipulated variables. Fortunately, most of these combinations are not useable due to various constraints, such as economics. From a steady-state degree of freedom analysis, there are only two degrees of freedom, since a total condenser is assumed. If the column had a partial condenser, there would, of course, be three degrees of freedom instead of two. Inventories that must be controlled are the reflux drum level (h_R), level in column base or reboiler (h_B), and the column pressure (vapor holdup). The remaining two variables are used to control the feed split and the fractionation.

The feed split is simply the amount of feed that leaves as distillate versus the amount that leaves as bottoms. The other variable, fractionation, is a reflection of the vapor and liquid traffic within the tower (or energy input). The overall column fractionation depends on the number of stages, the energy input, and the difficulty of separation. A control structure for this column (that follows a classical refinery distillation controls design) is shown in Figure 2.8.

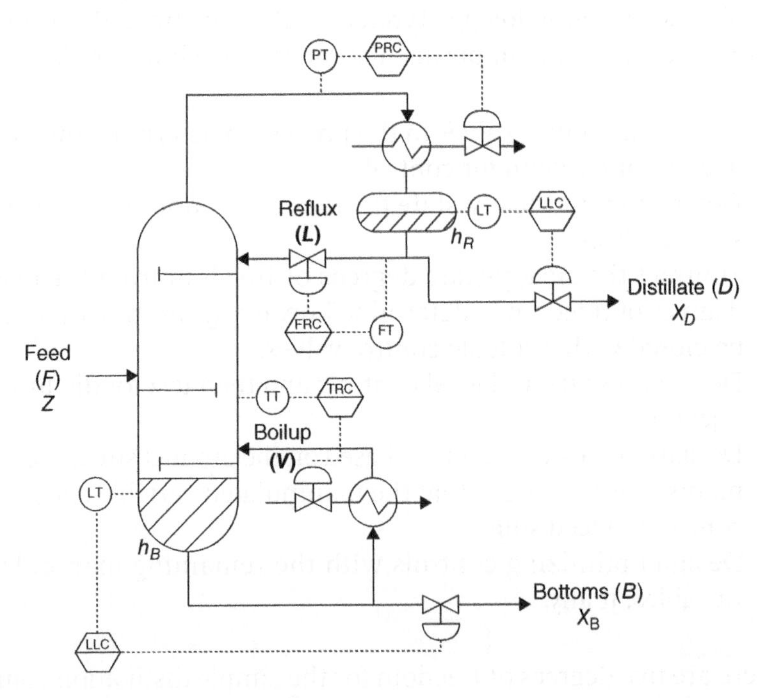

Figure 2.8 Column basic control scheme.

The most convenient method of verifying the operability of a proposed multivariable control scheme is through dynamic simulation. However, to effectively use dynamic simulation, it is first necessary to define the objectives of the control system, define the nature of the expected disturbances, and develop a basic understanding of the process both in terms of its steady-state and dynamic behavior.

2.5 Gain Analysis

From a practical point of view, the basic process design establishes the flow rate and composition for each stream and sets the mass and energy balances for the entire process. The preliminary control design usually occurs in the very early stages of the process design. Detailed design of the various components (e.g. vessels, drums, pumps, compressors) occurs much later and, generally, no further review or cross-check of the controls' design is performed. This very often results in poor control performance due to improper MV–CV pairing, particularly for inventory controls.

From a theoretical (and practical) point of view, the concept of the RGA provides engineers a quantitative comparison of how manipulating one stream will affect the relevant control variables. To elaborate further, when multiple, single loop control schemes interact, the closure of one loop can change the closed-loop gain of one or all the other control loops in the scheme. The SISO control loops may become unstable or respond sluggishly to disturbances since the overall loop gain has been altered (Eq. 2.6). The interaction between two control loops, in block diagram form, is illustrated in Figure 2.8 and is described as follows:

$$y_1 = a_{11}m_1 + a_{12}m_2 \tag{2.7}$$

$$y_2 = a_{21}m_1 + a_{22}m_2 \tag{2.8}$$

where:

y_i = the controlled or output variable "i"
m_j = the manipulated or input variable "j"
a_{ij} = the input/output relationship or transfer function between y_i and m_j

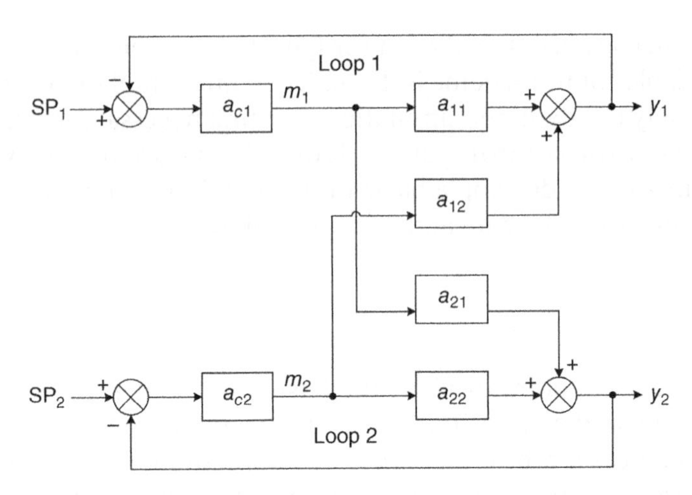

Figure 2.9 Loop interactions for a 2 × 2 system.

Figure 2.9 shows how a change in m_1 will affect both y_1 and y_2. There are two possible control configurations for a 2×2 interacting system such as this one. One could pair m_1 with y_1 and m_2 with y_2, or m_1 could be paired with y_2 and m_2 with y_1. The best control pairing is the one that has minimal interaction between the two control loops and remains stable across the entire operating range, rejecting load changes or random disturbances. For a control scheme containing n different controlled variables and n different manipulated variables, there are $n!$ possible control configurations. The relative gain array analysis, first introduced by Bristol (1966), offers a quantitative approach to the analysis of the interactions present between the required control loops, thus providing a method for appropriate pairing of MVs with CVs.

However, for the purposes of distillation control, a simplistic RGA is proposed to set up the base regulatory control.

Simplistic RGA (i.e. "Big Pipe v. Little Pipe"): If, in an existing plant, one needs to make a quick assessment of whether to use one stream over another for a particular CV, then a quick review of P&IDs or a field walk-down to determine pipe sizes makes this an effective method. This also works well if physical design information is not readily available. With this method, one is merely assessing, albeit on a gross basis, the relative effect of one MV versus the other; usually two flows, on a given CV (which, very often, is a vessel level). This method requires the least amount of information and, as implied in the description, requires only a visual assessment of pipe diameters to make the determination of which flow stream to use as the MV.

2.6 Common Control Configuration

A standard distillation column with a total condenser and distillate and bottoms streams has multiple combinations of control pairings that are possible. For this simple configuration, the following five (5) MVs exist:

1. Reboiler Duty (Q_R)
2. Reflux Flow (L)
3. Distillate Flow (D)
4. Bottoms Flow (B)
5. Condenser Duty (Q_C)

The feed and reflux flow rates are not considered as potential manipulated variables due to:

- The column feed stream is usually the MV of other units. Or restated, its characteristics are determined by the operating condition(s) of upstream units.
- As described previously, reflux is not an independent variable.

Similarly the following five key process or controlled variables (PVs or CVs) are identified as CVs that must be maintained in real time:

1. Condensor (Reflux Drum) Liquid Level
2. Reboiler (Bottoms) Liquid Level
3. Distillate Composition
4. Bottoms Composition
5. Column Pressure

Given the above list of CVs, the implicit operating objective may be described as "Double-Ended Composition Control" since both outlet streams' compositions are specified as control objectives. In decades past, this was not the case, but times have changed! This historical facet and the effect it had on distillation control will be explored further in Chapter 5.

From a practical point of view, in a standard distillation column configuration, the column pressure is expected to be controlled using the condensor duty, Q_C (further details are provided in Chapter 4). To further clarify: Q_C and Q_R must be equal for the energy balance to close; that is they are not independent! As such, there are four remaining inputs available to control the four outputs. Consequently, there are *4*! different possible control combinations as shown in Table 2.1.

Table 2.1 Possible pairings available for dual composition control.

Case	Reflux drum level	Bottoms level	Distillate composition	Bottom composition
1	D	L	B	Q_R
2	D	Q_R	B	L
3	L	D	B	Q_R
4	$\boldsymbol{L}$	$\boldsymbol{B}$	$\boldsymbol{D}$	$\boldsymbol{Q_R}$
5	B	L	D	Q_R
6	B	Q_R	D	L
7	Q_R	D	B	L
8	Q_R	B	D	L
9	D	L	Q_R	B
10	$\boldsymbol{D}$	$\boldsymbol{Q_R}$	$\boldsymbol{L}$	$\boldsymbol{B}$
11	L	D	Q_R	B
12	L	B	Q_R	D
13	B	L	Q_R	D
14	B	Q_R	L	D
15	Q_R	D	L	B
16	Q_R	B	L	D
17	D	B	Q_R	L
18	$\boldsymbol{D}$	$\boldsymbol{B}$	$\boldsymbol{L}$	$\boldsymbol{Q_R}$
19	B	D	L	Q_R
20	B	D	Q_R	L
21	L	Q_R	D	B
22	L	Q_R	B	D
23	Q_R	L	B	D
24	Q_R	L	D	B

While all of the above combinations are mathematically feasible solutions, only a small subset are typically found in actual implementations. The three common pairings, 4, 10, and 18, which are shown in boldface in Table 2.1, are the most common implementations due to the pairing rules described in Section 2.4. And to be clear, while the other pairings are not typically found in industrial settings, there can be situations in which *atypical* configurations provide the solution for unusual operating conditions. This topic will be covered further in Chapter 4 under "Relative Gain Analysis, aka Closing the Loop in Plant Design."

The typical pairings, 4, 10, and 18, are commonly referred to as Distillate and reboiler duty (DV), Reflux and Bottoms flow (LB), and Reflux and reboiler duty LV control configurations, respectively.

2.7 Screening Control Strategies via Steady-State Simulation

One option to evaluate which control structure(s) makes sense for a given specific distillation operation is to develop a steady-state simulation. The advantages of such an approach include:

1. It leverages the work done by the process designers by extending the use of the design steady-state simulations to control work.
2. The process control design can parallel the equipment design work and can identify where the limits inherent in the process design might result in controllability issues. This analysis can then be used to drive changes to the process design to expand or reduce the capacity of specific pieces of equipment; this facet will be expanded upon in later chapters.
3. It provides the beginning justifications for including various measurements and control valves that an over-exuberant project manager might try to eliminate to "reduce project costs." (A little recognized fact is that, for every grass-roots HPI project, the entirety of the control and instrumentation [C&I] budget, including engineering, instruments, the control system, and bulk materials, constitutes less than 2% of the entire project cost! Thus, even should the C&I costs double, it barely shows up in the "noise" for the overall project!).

The following steps summarize the approach:

1. Adjust the simulation specifications (flows, quality targets, etc.) and record the effect on base regulatory control loop objectives (temperatures, flows, pressures, etc.).
2. Run multiple steady-state simulations across wide ranges of operating scenarios.
3. Assess how the above adjustments affect the steady-state operating points, paying careful attention to incremental changes when moving between operating scenarios.

This approach provides a picture of the steady-state behavior across the expected operating envelope, which, in turn, identifies potential nonlinear relationships and the need for careful assessment of detailed equipment design, including vessel, control valves, heat exchangers, and piping.

Tutorial and Self-Study Questions

2.1 List an example of mass balance and energy-balance-based composition control?

2.2 If a distillation tower is operating at design capacity, which control scheme (mass or energy balance scheme) is likely to be an effective composition controller?

2.3 If you only have access to a steady-state simulation of the plant, can you still identify optimal controller pairing for a distillation tower?

2.4 Describe the overall distillation column control design procedure.

2.5 Describe direct and indirect distillation column control schemes.

2.6 Describe DV, LV, and LB distillation column control schemes.

2.7 Describe how cascade control might be useful in distillation column control schemes.

2.8 Describe how feedforward/ratio control might be useful in distillation column control schemes.

2.9 What are the ONLY TWO control "handles" (NOT Degrees of Freedom) on a simple distillation column?

2.10 Describe variable pairing.

2.11 Describe the relative gain array.

2.12 Describe how the relative gain array is used to select preferred variable pairings.
2.13 Describe multi-loop tuning.
2.14 Where is a relative gain analysis (RGA) most useful vis-à-vis distillation tower controls? What might it reveal/indicate? Why is this important?
2.15 What potential problem(s) is created by a distillation tower feed preheater that uses the subject tower's bottoms stream as the heat source?
2.16 What process stream condition or consideration determines the design pressure of a distillation column?

References

Bristol, E.H. (1966). On a new measure of interactions for multivariable process control. *IEEE Transactions on Automatic Control* **AC-11**: 133.

Buckley, P.S. (1964). *Techniques of Process Control*. New York: Wiley.

King, C.J. (1980). *Separation Processes*, 2e. New York: McGraw-Hill.

Tyreus, B.D. (1992). Selection of controller structure. In: *Practical Distillation Control* (ed. W.L. Luyben), 178–191. Van Nostrand Reinhold.

3 Control Hardware

The implementation and execution of the control structures discussed in this book requires dedicated control hardware, which generally is the domain of control systems engineers. Therefore, the objective of this chapter is to identify the nuances and intricacies of control hardware and how these aspects should be factored in the design process of distillation control systems. If the readers are interested in a more in-depth look at controller hardware, please refer to the books *Process/Industrial Instruments and Controls Handbook* (McMillan and Considine 1999) and *A Real-Time Approach to Process Control* (Svrcek et al. 2014).

3.1 Introduction

As the price of computing hardware has dropped, the means and capacity for providing more complex and CPU-demanding functions and features have expanded greatly. And control systems vendors have eagerly taken advantage of this. The great explosion of capabilities, however, has had no impact on the foundational requirements for performing robust and reliable controls:

1. Measurement (or inferring) of a process condition;
2. Calculation of the desired action (if any);
3. Actuation of some mechanical device to affect the process condition.

A Real-time Approach to Distillation Process Control, First edition. Brent R. Young, Michael A. Taube, and Isuru A. Udugama.

Companion website: www.wiley.com/go/Young/DistillationProcessControl

While the second item is the area most discussed in typical control course, which go into detail on various control equations and sometimes tuning of same, the first and third are just as important – perhaps more so – as, without these devices being properly sized and appropriately selected, the calculation effort is, at best, erroneous or, worse, useless. Thus, it is imperative that the prudent engineer ensures that all three items are thoroughly and properly selected and designed. Borrowing from the common metaphor of the three-legged stool: All three (3) legs must be capable of bearing the load equally; if any one leg is weak, then the stool (and its occupant) falls. Each of the three elements are addressed in the sections that follow.

3.2 Control Hardware Overview

From a hardware perspective, the implementation of process controls requires three main components.

1. Sensors: Sensors allow for monitoring the variables of interest in a distillation process. Pressure, temperature, flow, and level are the main variables in a distillation application, while direct composition and density measurements can also be made. Sensors generally capture information by correlating a change in electrical property to a change in the variable (e.g. changes in the electrical resistance of a metal component due to changes in temperature), such as in resistance temperature detectors (RTDs). This information is later converted to a digital signal that is used by the controller.
2. Controllers: Controllers calculate the next control move based on the information gathered from sensors, set-point, and other inputs. In distillation processes, the controller is generally a part of the distributed control system (DCS). A supervisory control and data acquisition (SCADA) system is an alternative architecture that is sometimes used in smaller installations. Programmable logic controllers (PLCs) are another option (McMillan and Considine 1999; Svrcek et al. 2014).
3. Final Control Element (FCE): FCEs receive the control moves information from the DCS and affect the physical process. Control valves are the most common FCE in distillation columns (McMillan and Considine 1999; Svrcek et al. 2014). However, a

reboiler heating element, pumps, and cooling fans are also FCEs related to distillation control.

Since the late 1970s, DCSs have been available as packaged offerings that contain both hardware and software solutions that together enables a centralized human–machine interface (HMI) to carryout process control in a plant (Dunn 2015; Williams and Kompass 1990). Due to the digital nature of the DCS, the process control actions are carried out in cycles, where sensor information from the current time is used to calculate the control move to be executed in the next cycle.

3.3 Sensors

When one contemplates the various measurements needed to accomplish the economic and operational objectives of a plant and its varied pieces of equipment, one usually thinks of flows, levels, temperature, pressure, and composition. But only with certain measurements – namely pressure – it is directly detected; all of the others are inferred and the type, quality, and reliability of the inference is determined by the type of mechanical device(s) employed vis-à-vis the thermodynamic conditions of the process fluid(s) and the desired rangeability and accuracy of the measurement.

As will be addressed in Chapter 3, a distillation process is governed by a mass and energy balance. The key objective of the sensors in the context of distillation control is to accurately capture dynamic changes to the mass and energy balance, which, in turn, enables actions to be performed to monitor or shift those conditions. In an ideal world, one would like to know the molar flow rate of components that are entering and exiting a distillation column in real time. Dynamic changes to the mass and energy balances are then calculated from this information, followed by the determination of the necessary control actions.

In reality, however, molar flow monitoring is prohibitively expensive and slow. Thus, a more practical approach is to monitor the mass or volumetric flow rate of each stream and estimate (or infer) the composition by proxy or other means when necessary (e.g. using temperature and pressure information together with an understanding of the vapor liquid equilibrium [VLE] to infer composition). This information can approximate the overall mass balance and dynamic deviations to it. Additional composition information at different tray locations of a column can also be captured in this way, enabling an accurate and precise estimation of the internal column dynamic behavior.

3.3.1 Process Considerations

"Rangeability" and accuracy are often not considered properly when selecting a measurement device, as many design engineers only consider "design conditions" or steady-state operations at the design conditions. This is a travesty and represents a failure to recognize that once the plant or equipment is put into service, it will only rarely be operating at the design conditions: it will spend far more time either turned-down or maxed-out!

The rangeability and accuracy of a sensor will dictate the accuracy of the information provided to the control systems and, thus, affect the accuracy of the plant's operations for that measurement. The rangeability of a sensor refers to the detection range that ensures nominally consistent accuracy. Ideally, the rangeability should cover the overall operating range of the process, including start up and shutdown (e.g. a level transmitter should ideally be able to monitor between 0% and 100% liquid level in a condenser or a reboiler drum). In practice, however, there may be practical limitations that prevent such a range being achieved (e.g. a typical-level transmitter that can only monitor between 25% and 100%). If rangeability is of critical importance, that is for safety or optimization, then alternative sensor technology with a broader rangeability can be used (e.g. switching to a level transmitter that can monitor 5–100% of level), whereas the accuracy of a sensor refers to its measurement error, which should ideally be zero, but, in practice, is determined by the type of sensor (e.g. if temperature is used as a proxy for estimating composition in a high-purity distillation column, then a higher level of accuracy may be required to accurately capture the change in composition (Luyben 1992)).

The location of the sensors is another important aspect that must be taken into consideration (e.g. a temperature sensor that is placed in the middle of the rectifying section may better capture a change to a distillate composition in a high-purity column application (Luyben 1992)). Placing a sensor in the vapor phase, as opposed to the liquid phase, will increase the general sensor dynamics, as changes in internal column traffic are first reflected as a change in vapor flow (and temperature). With this general information in mind, let's look at type of sensors that are used in distillation applications. Note: The term transmitter is used interchangeably with sensor in this chapter. This is because, sensors used in control "transmit" information to the controller. However, the physical reality is that the sensor and transmitter are separate devices and are often distinguished as such on engineering drawings (i.e. piping and instrumentation diagrams [P&ID]).

3.3.2 Flow Measurement Devices

The most common measurement – perhaps not in quantity, but of economic and operational interest – is flows. There are two fundamental types of flow measurement devices: volumetric and mass, with the former having the far larger variety of choice. Volumetric devices include orifice plates, venturis, vortex, roto, and mag meters.

Volumetric devices include orifice plates, venturis, vortex, roto, and mag meters. Even when the flow indication is shown with units of mass (on the control system screen), if the sensing device consists of any of the listed devices, then it is actually measuring (inferring) volume. The fluid density (which is needed to convert volume to mass) is predicated on an expected (i.e. design) temperature and, in the case of gases, pressure; more likely as not, the actual temperature and pressure will differ from the design. For liquid streams, the error resulting from any deviation from design conditions is usually small – for all practical purposes, negligible – but they should be confirmed nonetheless. For gases, however, the error resulting from deviation from design conditions can be significant. Thus, when utilizing volumetric devices, one must always consider the potential effects of deviation from design conditions.

Devices that directly measure mass flow (i.e. Coriolis meters) are, fortunately, insensitive to variations in fluid density. Early versions of these devices were quite bulky, especially as the pipe line size increased. However, continued development of the technology by multiple vendors has resulted in "drop-in" replacements that appear to the casual observer as merely a "spool piece" inserted into the process piping! This is extremely convenient for large-bore pipe applications (i.e. >6 inches or 152 mm). While more costly compared to "typica" volumetric devices, Coriolis meters have *none* of the turn-down limitations exhibited by many of the volumetric devices. This is a vital consideration when assessing operating scenarios that are well outside the design conditions.

In terms of the physical principles affecting the design and operations of many industrial volumetric flow measurements, many are based on the concept of differential pressure. Differential pressure measurements

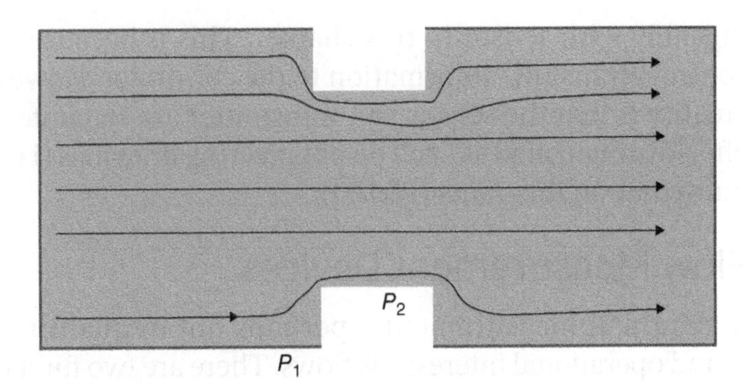

Figure 3.1 Orifice plate and differential pressure-based flow rate calculations.

require orifice plates or other flow constrictions. These devices work on Bernoulli's principle which states that an increase in the speed of fluid flow results in a corresponding drop in hydrostatic pressure. Figure 3.1 illustrates the principle of an orifice plate.

The orifice plate reduces the area through which the fluid can pass, resulting in an increased fluid velocity, which, in turn, results in a drop in local hydrostatic pressure. The volumetric flow rate is calculated based on the pipe diameter, the diameter of the orifice, as well as the pressure drop created by the orifice plate. In general terms, the following proportionally holds:

$$P_1 - P_2 \propto (\text{Flow velocity})^2 \propto (\text{Volume flow})^2 \tag{3.1}$$

Orifice plates have limited rangeability and require straight line pipes for accuracy and they are affected by changes in viscosity, density, and they create irrecoverable pressure loss in the line. However, orifice plate meters are relatively inexpensive to install and have a proven track record in industrial applications.

An alternative flow measurement that also works on the Bernoulli principle is a venturi flow meter. This device has the advantage of reducing the irrecoverable pressure loss (Considine 1995). However, they are more expensive due to being fabricated from a solid metal block which is precisely machined to the design specifications.

Electromagnetic flow meters, or "mag meters", operate based on Faraday's law of magnetic induction, providing a relatively accurate and linear signal. The principal benefit of Magnetic flow meters is that they do not disturb or restrict the flow (as do differential pressure-based measurements) and are not affected by change in fluid properties (other than conductivity). However, the cost of installation is higher. In addition to these technologies, ultrasound, vortex and turbine flow meters can also be employed to monitor volumetric flow in a pipe (Smith and Corripio 2006).

In recent years, Coriolis meters have become a popular, albeit more expensive, alternative to measure flow (Smith and Corripio 2006). In comparison to other technologies, the main benefit of a Coriolis meter is its ability to directly monitor the mass flow rate of a stream. This feature is very beneficial in establishing an accurate mass balance on a distillation operation. These flow meters work based on the Coriolis effect that directly correlates mass flow rate to the twisting or vibrational effect caused by the fluid on the flow tube(s) inside the meter. The main disadvantage of this system is its higher relative cost. Other advantages include accuracy at turned-down conditions and the direct measurement of density from the same device.

3.3.3 Pressure Measurement Devices

After flows, the next most common measurement (for control purposes) are inventory measurements: namely liquid levels in vessels and vapor space pressures. In both cases, ensuring that the mass balance accumulation is *Zero* is the ultimate control objective. Fortunately, pressure constitutes the basis for the vast majority of all measurement devices. Thus, it is the simplest and most well understood. The historical basis for the preponderance of pressure-based devices and the nuances of their accuracy and precision is beyond the scope of this book. Nevertheless, understanding the basis for these devices and, most importantly, their limitations is well worth the effort and the reader is encouraged to do so!

A distillation column's pressure can be measured using multiple technologies, including methods such as piezoresistive and capacitive sensors. For further details refer to Doran (1997) and DeHennis et al. (2016). From a process control perspective there are two varieties of pressure measurements:

- Absolute Pressure: This pressure measurement is based on the absolute pressure in a vessel. This measurement device has more stringent calibration requirements than other devices as it must be accurately calibrated to absolute zero (i.e. full vacuum). To this end, this type of a pressure measurement should only be used when absolutely necessary, such as in vacuum operations where precise measurement of pressure is warranted.

- Gauge Pressure: This measurement provides the pressure inside a vessel relative to atmospheric pressure. In distillation applications, many pressure transmitters are therefore in gauge pressure. It should be noted that "gauge pressure" is affected by changes in atmospheric pressure which is affected by altitude above mean sea level (MSL), as well as weather conditions.

3.3.4 Level Measurement Devices

In many level measuring devices, differential pressure is the basis for calculating the liquid level in a vessel. Differential pressure sensors (Considine 1993; McMillan and Considine 1999) work by using a diaphragm, where the two measurement points are connected to the opposite sides of the diaphragm. The deflection of the diaphragm indicates the differential pressure. In level sensing operations the differential pressure diaphragm will monitor the difference between the pressure at the bottom of the vessel and a reference pressure in the vapor space above the liquid. If the vessel is operating under "atmospheric conditions" (i.e. neither in vacuum service or pressurized), then the reference pressure port is open to the atmosphere. All other conditions require that the reference pressure is taken at the headspace of the vessel as detailed in Figure 3.2.

The level measurement is calculated using Eq. (3.2).

$$\Delta P = P_1 - P_2 = \rho g H \tag{3.2}$$

where ρ is the density of the liquid, g is acceleration due to gravity, H is the level in the vessel, and ΔP is the calculated differential pressure. To perform this calculation, one must know the density of the liquid so that the selected transmitter's calibration range covers the full operating range of the vessel's liquid inventory. To put a finer point on this: all commercially available transmitters have a maximum ΔP range, usually expressed in units of "Inches of Water Column." Thus, when specifying a transmitter for a particular-level measurement service, the fluid density must be known in order to ensure an appropriately ranged

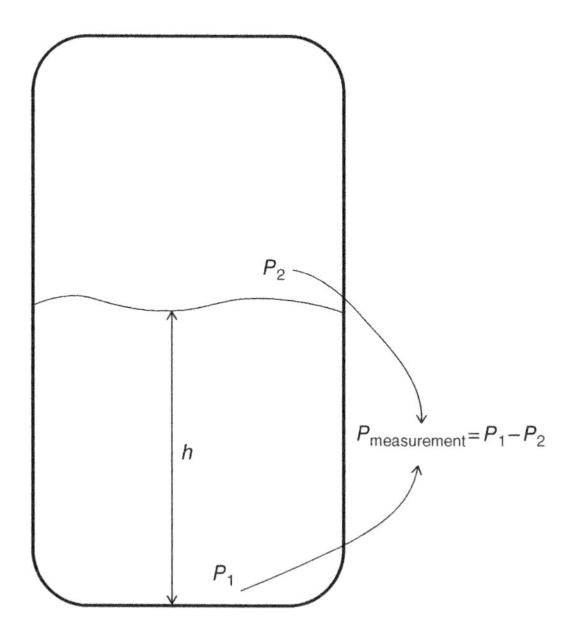

Figure 3.2 Schematic representation of how a level measurement can be derived using differential pressure.

transmitter is selected. Also be aware that any significant changes to the liquid density due to potential frothing or changes in composition will result in an incorrect-level measurement. "Frothing" has been the root cause of many process safety incidents wherein a level controller made improper control moves due to an incorrect-level measurement.

An alternative to measuring the level using differential pressure is to use a radar-type-level sensor (Fisher Educational Services Student Guide 1991; Svrcek et al. 2014). In this technology, electromagnetic waves are used to determine the level inside the vessel as shown in Figure 3.3. Stilling wells are often used for this type of device to reduce the noise in the level measurements, which is created by "sloshing." Radar-level devices are quite useful for dual-phase liquid measurements (i.e. when there is an aqueous and hydrocarbon layers in the vessel).

The device transmits electromagnetic pulses, which are reflected by the liquid surface. The time taken for the pulse to return is correlated to the liquid level in the vessel. The fact that the radar device does not "touch" the liquid can also be beneficial in hazardous environments. However, from a cost perspective, the differential pressure sensor will be significantly cheaper than a radar system. The challenge also often

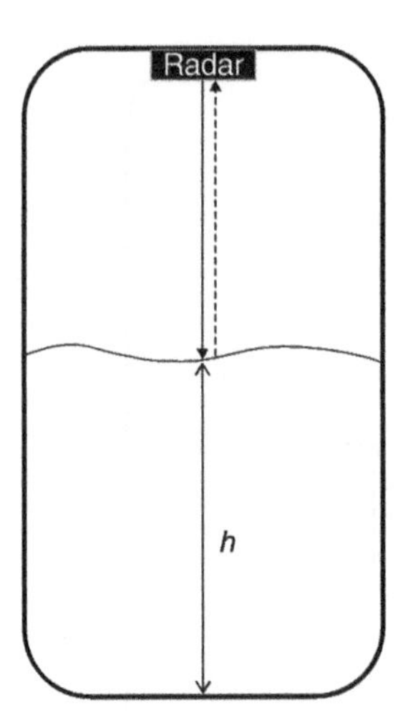

Figure 3.3 Schematic representation of how a level can be derived from radar measurements.

encountered with radar devices is in the initial setup and calibration. Therefore, the authors recommend that a qualified vendor representative be engaged to assist with the design and, especially, the installation and calibration of radar devices. This will save much grief during commissioning and startup.

Ultrasonic-level transmitters work similarly (Svrcek et al. 2014) to the radar system with the key difference being the use of sound waves instead of electromagnetic waves. Both the accuracy and rangeability of ultrasonic level measurements are lower than the radar devices. However, an ultrasonic-level transmitter is cheaper in comparison to radar.

Another technology, while quite popular in older facilities and still in use to this day is a displacer-style device, which is designed as a "bolt-on" self-contained instrument. It operates based on the principle of buoyancy and relies on a float or displacer within the meter's housing. As this device depends on internal mechanical linkages, it is not used as much as it was in previous decades. More recent displacer-style devices, while still relying upon an internal displacer, utilize an external sensing element rather than the internal linkages for determining the elevation of the float. As both of these technologies rely on a mechanical element, consideration must be given to the mechanical design of the vessels and the use of isolation valves, which allows maintenance to be performed on the device while the vessel is in service.

3.3.5 Temperature Measurement Devices

The last type of measurement device in wide use is the bi-metallic temperature sensor or "thermocouple" (Considine 1995; McMillan and Considine 1999; Svrcek et al. 2014). These devices rely on the Seebeck effect. While these are some of the most numerable measurements found in process facilities, due to their low cost, there are limitations to their applications due to the fact that they have relatively large fixed measurement ranges (based on the Thermocouple Type) and, due to the way that the measurement is detected, limited accuracy. The above notwithstanding temperature sensors are a cost-effective method to monitor and infer the composition of a distillation processes, as well as other temperature-sensitive operations (e.g. tube metal temperature in cracking furnaces).

The other types of temperature device is an RTD (Considine 1993; McMillan and Considine 1999; Svrcek et al. 2014), which monitors the change in resistance of a metal wire. Out of these two devices, the RTD has a much finer level of accuracy and rangeability. Moreover, the RTD provides a linear signal throughout its detection range (i.e. temperature is proportional to the signal). In contrast, a thermocouple typically has an "S" shape temperature vs signal relationship. In the past, the cost difference between thermocouples and RTDs was significant, with the RTD being the more costly option. Now, however, the cost difference between these two technologies is nominally insignificant. Given the negligible cost differences and the greater accuracy, the authors recommend the use of RTDs for every distillation temperature measurement. A common practice is to install an RTD AND a thermocouple for the same measurement: the former used for normal operations (and control), while the latter is used for startup conditions when the temperature is expected to be outside of the normal range for which the RTD is calibrated.

3.3.6 Direct Composition Measurements

In many distillation applications, the composition of a stream can be inferred using temperature. However, in some other applications, such as high-purity distillation, the direct measurement of composition is required, due to the poor correlation between temperature and composition. In these types of situations, gas chromatograph technology is often used to analyze the composition of a stream (Considine 1995;

Seborg et al. 2004; Udugama et al. 2017). The main constraint in using a gas chromatograph is the supporting infrastructure that is required. Unlike other measurement devices, chromatographs often must be housed in a dedicated equipment shed, which contains the sample system, as well as the chromatograph and its associated control system. While the cost of the GC itself is fairly reasonable, the cost for the design, installation, and configuration of the infrastructure (e.g. sample lines from the distillation unit, sample-handling tubing and conditioning system instrumentation and environmental control systems, as well as highly trained and dedicated maintenance personnel who ensure on-going accuracy and reliability) is what increases the cost of these systems. Nevertheless, in spite of the significant economic commitment, these systems provide a high level of return.

3.3.7 Maintenance

Maintenance is a vital factor in ensuring the accuracy and reliability of every measurement device. Failing to carryout routine maintenance often leads to sensor drift and/or failure. Inaccurate sensors often lead to incorrect conclusions about the state of a process. Hence, sensor maintenance is a pre-requisite to ensuring safe and on-specification operations. To this end, sensors should be located such they can be easily accessed for maintenance and qualified personnel be available to carryout routine maintenance and troubleshooting of these devices. The industry 4.0 drive is currently opening up the possibility for self-diagnosing sensors and predictive maintenance (Udugama et al. 2021), but the promise is still some time away from fulfillment. Even when the promise is realized, qualified human beings will still be needed to assess the information and then perform the required physical work to repair or replace the devices.

3.4 Final Control Elements

Final control elements (FCE) in distillation processes come in many forms. In most cases, however, the FCE in a distillation process is most likely to be a control valve. Control valves and FCE are a subject on their own and are covered in great detail in the *Process/Industrial Instruments and Controls Handbook* (McMillan and Considine 1999), while the *Fisher Control Valve Handbook* (Control Valve Handbook 2019) also covers this topic from an equipment and design perspective. This book, however, focuses on the influence of FCE selection for process control structures.

Many, if not most, control valves in industry use pneumatic (air) pressure to change the position of the valve, based on a signal sent by the control system. These devices are often provided with a valve position indicator, which clearly shows the actual valve position at any given point of time. This provides a visual "check" to ensure the valve is at the designated position.

3.4.1 Linearity

Nonlinearities in any part of the control structure represents an added complexity and FCEs, such as pumps and valves, can contribute to nonlinearities in the overall control loop.

Ideally, one should design the "system" (consisting of the pump(s), piping, and FCEs) such that the relationship between flow and FCE position exhibits linear characteristics throughout the FCE's operating range (i.e. a valve that exhibits a linear relationship between %valve opening vs flow rate). This is of particular importance if the valve position during normal operations, including startup and shutdown, needs to be set at high and low percentage openings. In such situations that the system exhibits nonlinearities, without any explicit linearization, the system will not be able to provide precise control at the edges of the operating range.

In such situations, the control signal from the control system should be mapped to a percentage valve opening that will provide a linear change in flow rate over the controller output signal range. Such a mapping means the nonlinearity caused by the valve is dealt with and the PID controller can be tuned without the need to consider this nonlinearity.

3.4.2 Time Constant and Failure Mode

It should be noted that FCEs can take time to move a given position or output. This is particularly true for large valves where the pneumatic system is responsible for moving the valve stem. While some texts attempt to illustrate and distinguish between all of the dynamic elements affecting the relationship between the measurement and the FCE, since very few of these elements can actually be changed once the process equipment and instrumentation is in place, one should simply recognize that they are all combined into a single dynamic response. Even so, wherever possible, every reasonable effort should be made to specify equipment that minimizes the dynamic characteristics of the system.

Many control valves may be specified with pneumatic positioners (Reinhardt 1997), which are used to ensure the valve moves into the commanded position. The specifics of valve positioner configuration are beyond the scope of this book, but there are many sources available, both hardcopy and online (e.g. YouTube), which may be referenced.

3.4.3 Mechanical Design Considerations

Lastly, one very important note about control valves: not every valve that is marketed as a "control valve" should be used for control applications. Specifically, certain mechanical styles of valves, namely butterfly and ball valves, fail to provide adequate effective control of the process. This is due to their inherent mechanical designs, which gives them an effective range of only 10–30% of stem movement. Outside of that range, they have NO CONTROL whatsoever on the process.

In short: NEVER use a butterfly or ball valve for continuous control applications! They are only suitable for On/Off or Open/Close functions.

3.5 Controllers/CPU

Many sensors use fundamental principles to translate a physical measurement about a chemical process into an electrical signal, be it a drop in resistance or a change in voltage. This information now needs to be communicated effectively to the control system, the most ubiquitous of which is the DCS. The DCS employs a network-type architecture and consists of multiple layers.

The objective of such a system is to localize control actions, which in turn improves the overall reliability of the system (Cobb 1996). All components in a DCS are made for extreme reliability, as failure (in particular of the bottom two layers) can effectively result in a loss of control. In a unit operation such as a distillation column, this can have catastrophic consequences. Independent safety control loops are installed to safely shut down a distillation process in such events.

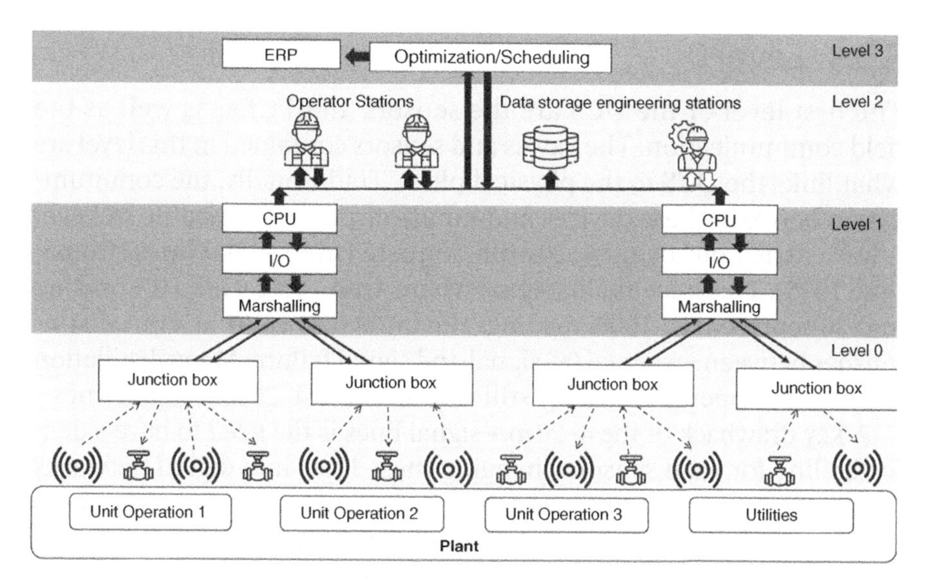

Figure 3.4 Hierarchical representation of a distributed control system.

These are often called safety instrumented systems (SIS) and are separate and independent of the DCS. Figure 3.4 shows the different levels of process controls and data management that can take place in a DCS system.

As illustrated in the Figure 3.4, a DCS consists of multiple levels of execution.

- Level 0: Consists of the sensors, FCE, and the field communication, which includes junction boxes, where multiple signal wires are collected and routed.
- Level 1: Consists of signal marshaling, I/O modules, and controls. In older installation, the signal handling and controls are two distinct levels, but in modern installations these are increasingly merged. Generally, regulatory and some advanced regulatory control will be performed at this level.
- Level 2: Consists of operator stations, data storage, and network switching. Multiple-Level 1 I/O and control module groups are combined at this level and provides a true plant-wide picture.
- Level 3: Consists of process optimization and scheduling layer. Level 3 also interacts with the enterprise resource planning (ERP) systems.

3.5.1 Level 0

The first level of the DCS are the sensors and FCEs, as well as the field communication. The valves and sensors contained in this level are what links the DCS to the physical plant. Traditionally, the communication between these devices and input/output (I/O) module in Level 1 was carried out using 4–20 milli-Ampere (mA) signal lines (Considine 1995). This is an analog signal where 4 mA represents a 0% reading and 20 represents a 100% reading. Having a zero value at 4 mA distinguishes between an actual 0% signal and signal failure. Many distillation processes in operations today still rely on these 4–20 mA signal lines.

A key drawback of the 4–20 mA signal lines is the need to have a dedicated line for each sensor. Changing these lines into digital networks greatly improves the information that can be sent back and forth between the plant and the DCS. However, replacing these lines with a full digital solution is not always economic. Highway Addressable Remote Transducer (HART) Protocol addressed this need by using the 4–20 mA signal lines for bidirectional communication with multiple sensors and control valves (Cobb 1996). At its core, HART uses a similar concept to old copper telephone lines, where a digital signal is transmitted using analog current. Many sensors and control valves sold in the market today are HART enabled.

In a modern plant, the designers also have the ability to opt for a fully digital network-type solution by choosing a protocol that is generally referred to as Fieldbus (ICE 2019). The Fieldbus communication protocols (profibus, modbus, etc.) are regulated by the International Electrotechnical Commission (IEC) 61784/61158 standards (2019). Simply put these protocols are somewhat similar to an Ethernet cable where a signal cable can be networked to multiple sensors and valves.

For efficient handling, multiple analog and digital loops are combined in a junction box. For 4–20 mA communications (including HART), wires from the sensors are combined into a single multi-core (multiple wire) isolated cable that is then sent through to the marshalling cabinets in Level 1. For Fieldbus-type implementations, a single (or multiple) Ethernet-type cables can carry out this task.

3.5.2 Level 1

Information from Level 0 is received by the marshalling stations (also referred to as the marshalling cabinet), which systematically sorts the signals. These signals are passed to the I/O modules where they are terminated. Both analog and digital signals can be received by the I/O modules. The I/O modules then pass this information to the controller CPU for further processing. The CPU will then carryout any controller calculations needed to decide the next controller move (e.g. based on the algorithm). To avoid any confusion between the controller and control hardware, we call the control hardware the CPU. The CPU consists of multiple cores and dedicated memory.

The key difference between this CPU and a domestic PC setup is the level of redundancies that are in place and the software that is focused on ensuring extreme reliability. Each I/O module and CPU will likely focus on signals to and from a single unit operation or a plant area. Assigning related sensors and valves to the same controller/CPU will enable a reduction in the need for Level 2 communication. However, for some advanced control configurations, this might be an unavoidable issue. The CPU will also handle the distribution of signals.

3.5.3 Levels 2 and 3

Level 2 consists of the operator station, engineering stations, and data storage links. Information from multiple Level 1 CPUs is connected via digital protocols to Level 2 infrastructure. For example, each operator station may draw inputs from multiple CPUs and issue commands to CPUs. The human–machine interface (HMI), which is a graphical representation of the current state of operations in the plant, is a key Level 2 asset. In fact, the HMI represents the only continuous point of contact between the human operators and plant operations. The operator will constantly monitor the performance of the plant based on the signals that were gathered at Level 0 and passed on to the HMI through Level 1. When needed the operator will issue instructions (such as set-point changes) to the controllers at Level 1 to achieve an overall operational objective.

The operator station in today's process plants are a set of computer/TV display screens and other processing hardware that display the current state of plant. Currently, many DCS vendors have standard graphics libraries, which can be referenced and used as a basis for building up an HMI. The International Society of Automation (ISA) 101 committee is currently developing a set of general standards, recommended practices, and/or technical reports that can be followed when developing HMIs in manufacturing applications.

Engineering stations are terminals that are used by process control engineers and control systems engineers to carryout engineering changes to the control systems. Tasks that can be performed in such a terminal include changes to alarm limits as well as some changes to control structure. In practice, using different login credentials on an operator HMI can open up an engineering station.

Level 3 is a layer where optimization tasks such as model-based control can be carried out. In addition, scheduling tasks that are related to the overall production process are also carried out in this layer. Level 3.5 is commonly referred to as a layer that interacts between the plant operations and the business layer that may consists of ERP software.

3.5.4 General Set Up and Considerations

In terms of setting up a DCS, a process control engineer would mainly be involved in setting up the control loops, while the "physical" wiring and information management will be carried out by control systems engineer. In theory, the generalized communications protocols used between and with-in the layers mean devices from multiple vendors can be "mixed and matched" to create a DCS. However, in reality the complexities in integration means many plants will use a single vendor for their DCS solution in Levels 1–2. Level 0 is somewhat vendor neutral as valves and transmitters from multiple vendors can be integrated to the DCS system with minimum effort. In Level 3, a handful of specialized service providers as well as DCS vendors provide solutions.

It should also be noted that the implementation of control structures is done on software tools that are provided by the vendors. The current generation of these software tools looks somewhat similar to the way an industrial process simulator can be configured, albeit, with further details being required about aspects such as signal treatment, function blocks, and execution order.

3.6 Modern Trends

Digitalization, Internet of Things (IoT), and Industry 4.0 are mega trends that are affecting management decisions in many manufacturing industries (Udugama et al. 2022). Most of these concepts benefit from the tried and tested DCS technology as it is a key enabler for implementing these concepts within the technical and safety constraints of industrial processes. At its core, the DCS is a localized cyber-physical system. In the mid-1900s, control systems went through a similar transition where pneumatic communication systems were replaced by electrical communication systems (4–20 mA lines). This transition significantly expanded the capabilities of controllers and sensors. Today, we are at the verge of a similar transition.

3.6.1 Wireless Communication and Smart Devices

Currently many valves and sensors in industrial distillation processes are connected through 4–20 mA analog wires. However, in other manufacturing industries, wireless sensors are used in many control applications. Implementing wirelessly communicating valves and sensors would significantly reduce the need for wiring and signal management. This is an attractive proposition, in particular during plant revamps. With concepts such as the IoT, there is also the possibility to install "smart" sensors and valves that can communicate with each other and calculate and execute relevant control actions at least at the regulatory control level.

Unlike in other industries, process safety is a paramount concern in distillation processes. As such, a key hurdle in implementing these types of wireless systems and smart controllers are the cybersecurity threats, e.g. terrorist hijacking a communication to a critical valve, which can lead to a catastrophic event leading to loss of life. At the same time, extreme level of reliability of communication between the CPU and these devices (or communication between devices) must also be guaranteed. A path forward is to initially implement such systems on non-safety critical loops and build up operational experience and expertise, which will then enable implementation of these concepts on safety critical systems.

3.6.2 Smart CPUs

Traditionally the CPU used for process controls (Layer 1) have been hardcoded systems using custom operating systems and custom logic. With the limited availability of computing power, these CPUs were optimized to carry out the key tasks at hand, which are signal processing and PID execution. However, with increased processing power that is available and the introduction of higher-level programming languages, these CPUs are now capable of carrying out complex optimization activities on the same hardware (albeit, on separate CPU cores and memory). These capabilities reduce the hardware barriers that existed to implement optimization-based controls including model predictive control.

3.6.3 Digital Twins

Closed-loop control of a complex process based on model prediction represents an advanced form of digital twins (e.g. Yu et al. (2022) and Udugama et al. (2021)). With the current advances in computing power implementation of digital twins as dedicated process, optimization tools can be considered. These types of digital twins can interact with the DCS at Level 3.

Tutorial and Self-Study Questions

3.1 What is the difference between direct and indirect measurement?
3.2 What is measurement accuracy and how is it reported?
3.3 What do the following terms mean? Repeatability, precision, sensitivity, gain, attenuation, dead-band, resolution, hysteresis?
3.4 What are the types and sources of measurement error, and how can one analyze, accumulate, and quantify them?
3.5 What are some of the basic components of typical measurement systems?
3.6 How might measurement systems be calibrated?
3.7 What types of signal transmission methods are used in process plant?
3.8 What are primary elements, sensors, and transmitters? What are the primary types of measurement used in process plants?
3.9 What are the common static types of pressure measurement and how do they work?
3.10 How is pressure measured dynamically for continuous automatic control purposes? How do these pressure measurement devices work?
3.11 How is level measured and, in particular, how is it measured dynamically for continuous control?

3.12 What are the two major types of temperature measurement, broadly, how do they work, and what are their relative merits?
3.13 What are two major methods of measuring flow, fundamentally how do they work, and what are their pros and cons?
3.14 What are the primary types of final control elements that are used in process plants and why?
3.15 What are the primary types of control valves and their characteristics?
3.16 What considerations need to be made when specifying a control valve?
3.17 What is the control valve coefficient and how is it calculated?
3.18 What is the basic valve sizing procedure?
3.19 What is a valve positioner and why would one be used?
3.20 Broadly, how does one account for control valve dynamics in process plants?
3.21 What is a DCS?
3.22 What is a PLC?
3.23 What are the relative merits of DCSs and PLCs? How do you think this will change over the next decade?
3.24 How are industrial control algorithms, to avoid excessive derivative, kick on set-point changes?
3.25 What is reset windup and how is it avoided in industrial controller algorithms?
3.26 What is the difference between Percent-of-Scale (PoS) versus Engineering Unit (EU)-based control systems? Why is it important to know the difference? How does this affect transmitter range changes?
3.27 What Percentage of a Greenfield Capital Project budget does the entirety of Control & Instrumentation (C&I) consume (including bulk materials, engineering, instrumentation, DCS)? Why is this and what effect does this have on attention given to C&I work?

References

Cobb, J. (1996). Control in the field with HART® communications. *ISA Transactions* 35 (2): 165–168.

Considine, D.M. (1993). *Process/Industrial Instruments and Controls Handbook*, 4e. New York: McGraw Hill.

Considine, D.M. (1995). *Industrial Sensors and Measurements Handbook*. New York: McGraw-Hill.
Emerson Fisher (2019). *Control Valve Handbook*, 5e. Fisher Controls International LLC.
DeHennis, A., Chae, J., and Baroutaji, A. (2016). *Pressure Sensors, Reference Module in Materials Science and Materials Engineering*. Elsevier.
Doran, C. (1997). *Matching pressure transmitters*. Fluid Flow Annual: *Chemical processing*.
Dunn, T. (2015). *Basics of control systems*. In: *Manufacturing Flexible Packaging*, Chapter 10, 103–110. William Andrew Publishing, Elsevier.
Fisher Educational Services Student Guide (1991). *Fundamentals of Level Measurement*. USA: Fisher Controls International Inc.
ICE (2019). Industrial communication networks – Fieldbus specifications – INTERNATIONAL ELECTROTECHNICAL COMMISSION, IEC 61158-1 Edition 2.0 2019-04
Luyben, W.L. (ed.) (1992). *Practical Distillation Control*. Van Nostrand Rheinhold.
McMillan, G.K. and Considine, D.M. (1999). *Process/Industrial Instruments and Controls Handbook*, 5e. New York: McGraw Hill.
Reinhardt, N.F. (1997). Impact of control loop performance on process profitability. *Aspen World '97*, Boston, MA (15 October 1997).
Seborg, D.E., Edgar, T.F., and Mellichamp, D.A. (2004). *Process Dynamics and Control*, 2e. Wiley.
Smith, C.A. and Corripio, A.B. (2006). *Principles and Practice of Automatic Process Control*, 3e. Wiley.
Svrcek, W.Y., Mahoney, D.P., and Young, B.R. (2014). *A Real Time Approach to Process Control*, 3e. Wiley, Chapter 2.
Udugama, I.A., Wolfenstetter, F., Kirkpatrick, R. et al. (2017). A comparison of a novel robust decentralised control strategy and MPC for industrial high purity, high recovery, multicomponent distillation. *ISA Transactions* 69: 222–233.
Udugama, I., Öner, M., Lopez, P.C. et al. (2021). Towards Digitalization in Bio-Manufacturing Operations: A Survey on Application of Big Data and Digital Twin Concepts in Denmark. *Frontiers in Chemical Engineering* 3, 16 September 2021: 1–14.
Udugama, I.A., Lopez, P.C., Gargalo, C.L. et al. (2021). Digital Twin in biomanufacturing: challenges and opportunities towards its implementation. *Systems Microbiology and Biomanufacturing* 1: 257–274.
Udugama, I.A., Bayer, C., Baroutian, S. et al. (2022). Digitalisation in chemical engineering: industrial needs, academic best practice, and curriculum limitations. *Education for Chemical Engineers* 39: 94–107.

Williams, T.J. and Kompass, E.J. (1990). Distributed digital industrial control systems: accomplishments and dreams. *IFAC Proceedings Volumes* 23 (8), Part 4: 95–109.

Yu, W., Patros, P., Young, B.R. et al. (2022). Energy digital twin technology for industrial energy management: classification, challenges and future. *Renewable and Sustainable Energy Reviews* 16: 112407.

4 Inventory Control

In Chapter 2, we explored the concept of MV–CV Pairing as well as the concept of the relative gain analysis. In Chapter 3, we explored the hardware that makes the control concepts happen. With this understanding in mind, Chapter 4 will explore the role of inventory control in distillation operations in detail. In short, the proper management of inventory (both liquid and vapor) in a distillation column holds the key to predictable and stable operations. Thus, from a practical point of view, having stable and robust inventory control of a distillation column (or a dynamic process simulation) is a pre-requisite to establishing reliable composition control and, subsequent, optimization.

4.1 Pressure Control

Controlling the pressure inside a distillation column is paramount to ensuring stable vapor–liquid equilibrium, which then results in robust composition control. The means to control pressure in any distillation column typically consists of three main configurations, depending on the properties of the overhead vapor mixture and the temperatures of the available heating and cooling mediums. Prior to describing the types of configurations that can be implemented, it is important to note the following:

- In practice, the column pressure is measured in the tower's overhead vapor stream in between the condenser and the top of the column.

A Real-time Approach to Distillation Process Control, First edition. Brent R. Young, Michael A. Taube, and Isuru A. Udugama.

Companion website: www.wiley.com/go/Young/DistillationProcessControl

- If the overall column pressure profile is of practical interest (often as a constraint to maximizing feed and/or improved separation), then the differential pressure in the column must also be measured. This measurement should be made with a dedicated differential pressure meter rather than two separate pressure transmitters in order to minimize the effect of calibration errors and drift.

One other important item to note is that the operating pressure of the tower is dependent upon one important factor: the temperature of the reboiler heating medium. The explanation for this fact is left as an exercise for the reader.

4.1.1 Total Condenser

The most ubiquitous configuration that is used to control the pressure inside a distillation column is the total condenser configuration. In this configuration, the pressure is controlled by totally condensing all of the vapor leaving the top of the column into liquid. Condensing any given mixture from vapor to liquid results in a significant volume reduction, which in turn in the confines of a distillation column results in a reduction of pressure.

Configuring a total condenser to perform pressure control can be carried out in a few different forms. A conceptually straightforward configuration is displayed in Figure 4.1.

In this configuration, the pressure at the overhead of the column is monitored and used to manipulate the flow rate of cooling water (or other medium) that is used to condense the vapor stream. To this end, an increase/increase relationship can be observed (i.e. if the pressure is above set point the cooling medium flow rate should be increased). Also note that, in practice, throttling or modulating a cooling water stream is not recommended as this affects the fouling rate on the cooling water side of the exchanger, thus shortening the exchanger's service time.

A variation of this total condenser configuration that is often found in industry is the flooded condenser arrangement as illustrated in Figure 4.2. In this configuration, a shell and tube heat exchanger is used as the condenser: the overhead vapor enters on the shell side while cooling water is circulated on the tube side. As vapor on the shell side of the heat exchanger starts to condense into liquid, a liquid inventory builds up. As the inventory is allowed to increase, the tubes become

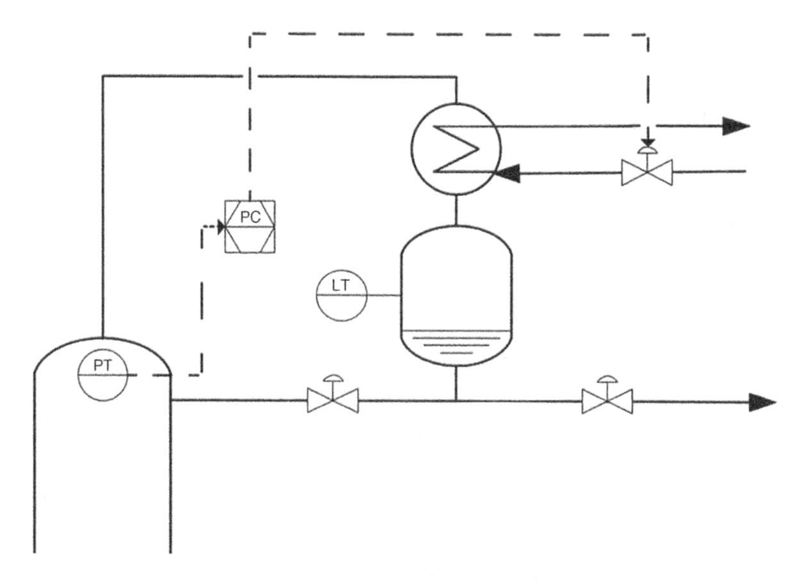

Figure 4.1 Basic total condenser-based pressure control.

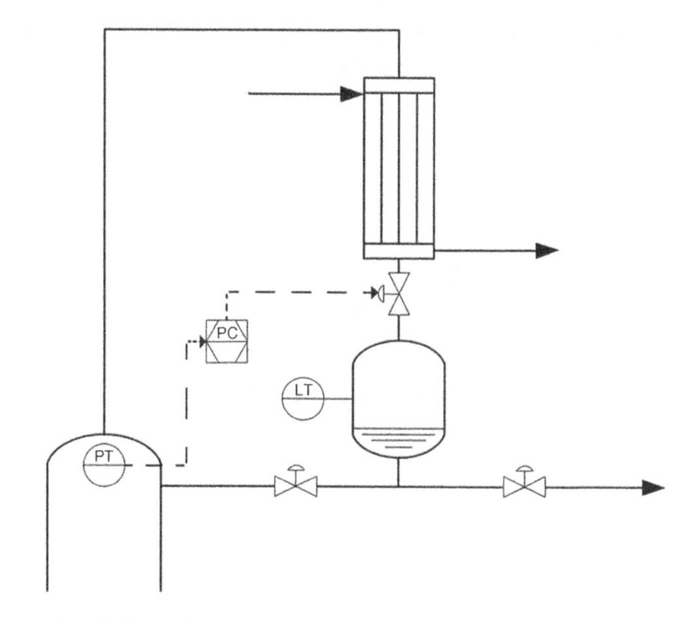

Figure 4.2 Flooded condenser arrangement.

submerged in liquid, thus reducing the area available for condensation. Manipulating the liquid flow rate out of the heat exchanger directly changes the heat transfer area available for condensing and, hence, the distillation tower's pressure. In practice, a cascade control scheme is sometimes configured where the pressure controller is manipulating the set-point of a level controller that is then manipulating the liquid flow rate out of the condenser. More often, however, the tower pressure controller directly manipulates the condensate flow valve. The simpler (non-cascade) scheme is quite stable and robust, as described in the following section.

4.1.2 Flooded Condensers

From a practical point of view, many refining and chemicals units' operations in industry makes use of the flooded condenser arrangement. Although flooded condensers are widely used, with a long and successful history, their physical and controls design present a few challenges. Before addressing the physical design aspects, let's first review *how* a flooded condenser operates by conducting a thought experiment using a typical design, as illustrated in Figure 4.3.

The thought experiment will take the form of an open-loop step test, as one might perform to tune the pressure controller on the distillation tower. Before getting into the thought experiment, there are a few items

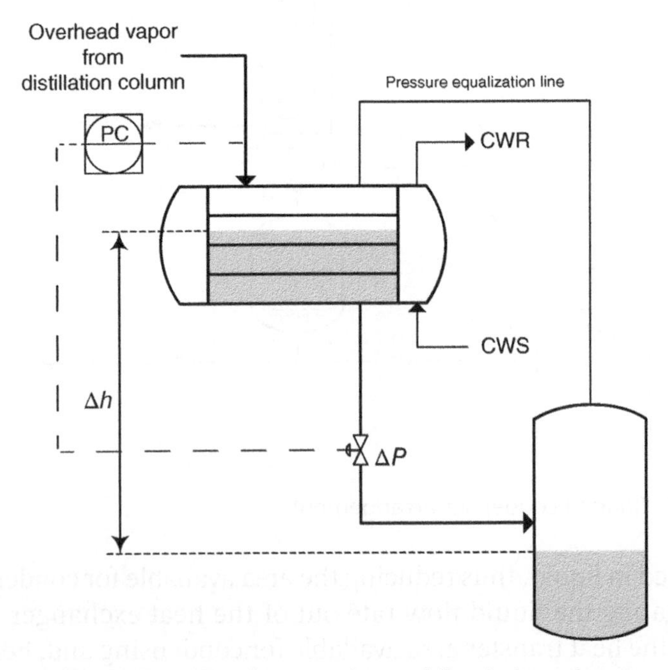

Figure 4.3 Diagram illustrating how an elevated flooded condenser operates.

to note regarding the sketch in Figure 4.6 and associated assumptions (which also apply to most real-world installations):

1. The condenser is at a higher elevation compared to the reflux drum. As shown in the sketch, there is a relatively small pressure equalization line between the condenser and the reflux drum that results in both vessels being at the same pressure, thus ensuring that gravity is the only driving force for moving the liquid from the condenser into the drum. To be clear, there is no vapor flowing within this equalization line, except for those transient events when the drum level is moving.
2. While the sketch has a shell-and-tube exchanger illustrated, the principles described below apply exactly the same for an air-cooled/finned fan exchanger.
3. It should be understood that there is *total condensation* of the incoming vapor to the condenser and that the exposed tube area is sufficient to accomplish same.

Beginning at steady state, all the forces and mass flows are in balance: vapor and liquid mass flow in/out of the condenser and the hydraulic pressure losses across the control valve (ΔP) versus the liquid head between condenser and reflux drum (Δh) all balance. When the control valve opening is increased, the following outcomes result:

- More liquid is allowed to flow from the condenser into the drum, resulting in the condenser liquid level beginning to drop, exposing more tube surface area.
- This, in turn, causes more vapor to condense, thus resulting in lowering the pressure in the condenser.
- The lower pressure in the condenser vapor space now propagates back toward the distillation tower reboiler allowing more vapor to be generated (with no required change in heat input to the reboiler), thus increasing the vapor flow rate through the tower into the condenser. Since the vapor velocity in distillation towers is very fast, this back-propagation of the lowered pressure and the increase in liquid vaporization from the reboiler occurs nearly instantaneously.
- Back in the condenser, the decrease in pressure drop across the control valve (due to the larger valve opening) balances out with

the decrease in liquid head between condenser and reflux drum, resulting in the condenser liquid level to stop dropping.

- The increased vapor flow entering the condenser keeps pace with the higher condensate flow and the increased condensation duty from the newly exposed tube area (e.g. maintaining total condensation), keeping the pressure at its new lower value.

With all the forces now back in balance, the condenser is at a new steady-state condition: a lower liquid level, a lower tower pressure, and higher condensate flow to the accumulator.

Another hotly debated topic concerning flooded condensers involves the use of a hot-vapor bypass (HVB). As illustrated in Figure 4.4, (hot) vapor from the distillation tower overhead line is diverted directly to the overhead distillate accumulator vessel. This stream is modulated using a control valve, which is used to maintain pressure differential between the tower and the accumulator. Then a second pressure controller maintains the accumulator pressure via a vent stream. This was (and, often, continues) to be a very common design, particularly in oil refineries.

However, while this was a very common design practice for refineries built decades ago, the real justification behind the need for this

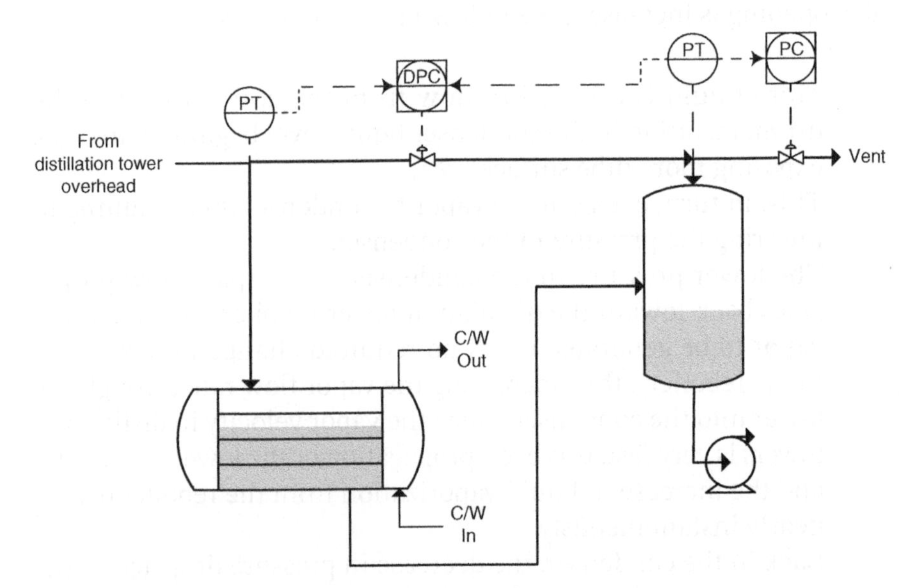

Figure 4.4 Hot-vapor bypass design on flooded condensers.

design has been long forgotten, to wit: In order to provide easy access for maintenance work, the condenser was located at grade (i.e. ground level). Furthermore, the accumulator was then located at a higher elevation. Thus, a pressure differential (ΔP) is required to provide the driving force that moves the condensate from the condenser to the accumulator, rather than a pump (which increased both capital and maintenance costs).

As illustrated in Figure 4.4, the differential pressure controller's (DPC) function is to maintain the requisite driving force that overcomes the elevation different plus the frictional pressure loss of the liquid flowing through the piping. This was a reasonable design choice given the capabilities and requirements of that era.

As maintenance practices and equipment capabilities improved and changed over time, the need to locate the relatively large exchangers at grade diminished. Thus, newer refineries located the condenser(s) at a higher elevation from the accumulator(s) (also resulting in a smaller "foot print" for the facility). However, the HVB design continued to be utilized even though the need for it had been eliminated. As a result, numerous HVB controls were abandoned and/or revised to allow the tower and accumulator to operate stably and reliably. While HVB designs can be successfully implemented today, there are too many nuances in the equipment's mechanical/physical design, which, if missed, will result in unstable performance. The design criteria described in Section 4.1.2 is wholly sufficient to address the needs of most, if not all, distillation tower's needs. Therefore, the authors have one recommendation regarding using HVB designs: DON'T!

Bottom Line: Keep it simple!

4.1.3 Sub-Cooled Reflux

A consequence of flooded condensers for both designs (described above) is sub-cooled reflux. That is, the condensed liquid becomes cooled below the saturation temperature due to the liquid volume held within the condenser and exposed to on-going heat transfer with the cooling medium. (Also note that subcooling may also occur even if the condenser is not flooded, as a result of extended contact time within the condenser between the condensed liquid and the cooling medium.) While the subcooling provides some benefit to the reflux pump NPSH, the *most significant* effect occurs when this sub-cooled liquid enters back into the column: an incremental amount of vapor coming up into the top tray is condensed on that tray, resulting in more than just the raw external reflux (ER) liquid flow that enters the tower going down within the tower. This phenomenon is commonly referred to as *internal reflux* (IR) and the thermal effect can be quite significant to the tower's behavior when the subcooling becomes "large." The effect is observed by the amount of reboiler duty required to maintain energy balance on the column: it will be larger than that required to vaporize the raw external reflux and distillate flows.

It should be noted that there are two (2) definitions for "Internal Reflux" that are commonly used, depending on the specific tower configuration. The alternative definition to the one described above is most often encountered with towers that have a major side-draw stream in between the tower feed and the reflux streams. So internal reflux, in that case, refers to the liquid that continues down the tower below the side draw. That is, the external reflux flow value minus the side-draw flow value. A typical example of where this definition applies is with a C2 Splitter in an ethylene plant. Even under this situation, however, subcooled reflux has the same effects, as described above, and these effects must be taken into account.

The presence of subcooling is found in many chemical and petrochemical plants and, while not found as frequently in refineries in years past, has become an increasingly frequent occurrence. However, recognizing the need to compensate for this phenomenon, and how to do so, in refineries seems to be lacking. Failing to explicitly account for subcooling will result in "unexplained" changes in stream composition(s) that then requires reliance upon feedback control to make necessary process adjustments in order to maintain quality measurements at their targets. This feedback correction may require "significant" time, depending on the dynamic response of the quality measure(s) versus changes in controller MVs. Implementing IR control eliminates the "unexplained" composition changes (caused by changes to subcooling), thus resulting in better control and less product quality variability, which is the ultimate goal of the tower's controls. IR is easily calculated and can be implemented in a control loop that explicitly accounts for the effects; such a controller is a very effective type of feed forward controller.

The formula for calculating IR is given in Eq. (4.1) and can be easily implemented in most control systems. Furthermore, little, if any, dynamic compensation is required for any of the inputs and there is no "tuning" required for an IR controller: it merely adjusts the external reflux by the difference in the tower overhead vapor temperature and the external reflux temperature (ΔT).

$$\mathrm{IR} = \mathrm{ER} \times \left(1 + \frac{\Delta T \times C_P}{\lambda}\right) \tag{4.1}$$

where:

C_p = the heat capacity of the liquid;

ΔT = the amount of subcooling (the overhead vapor temperature leaving the tower and going to the condenser minus the reflux temperature);

ER = the external reflux value; and

λ = the heat of vaporization (condensation) of the overhead fluid.

4.1.4 Partial Condenser

In a situation where a non-condensable fraction of the overhead stream is very small (trace non-condensable gasses), a similar control configuration to the total condenser pressure regulation can be used with the addition of a small (unregulated) vent stream affixed to the reflux drum, as illustrated in Figure 4.5. The objective of the vent in this instance is to consistently purge the trace non-condensable gases that would otherwise accumulate overtime in the reflux drum. Note: This solution still relies on the condenser cooling duty to regulate the pressure in the column while the vent is only used to ensure that there is no long-term accumulation of noncondensable gases that gets "trapped" in the reflux drum resulting in a pressure build up.

However, if the incoming overhead stream consists of significant levels of non-condensable gases, the control scheme illustrated in Figure 4.6 is recommended. In this scenario, the inventory of non-condensable gases in the condenser is regulated to achieve desired pressure.

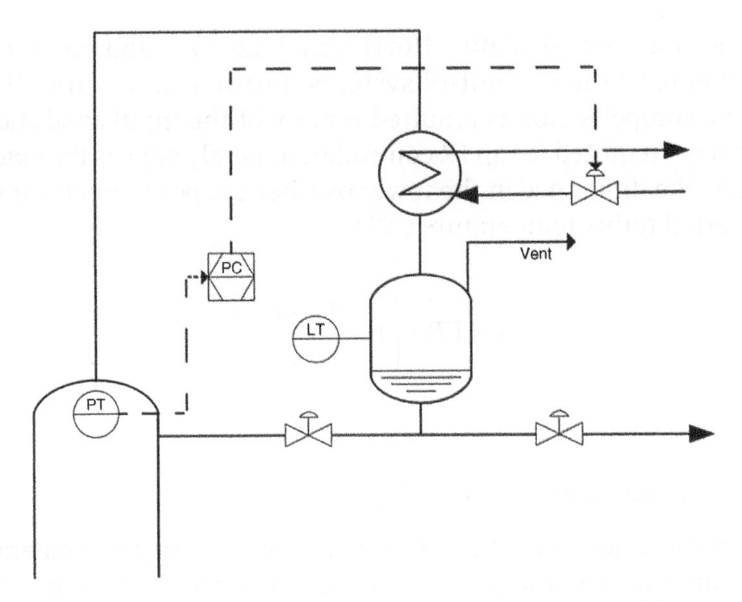

Figure 4.5 Pressure control with no or only trace non-condensable gases.

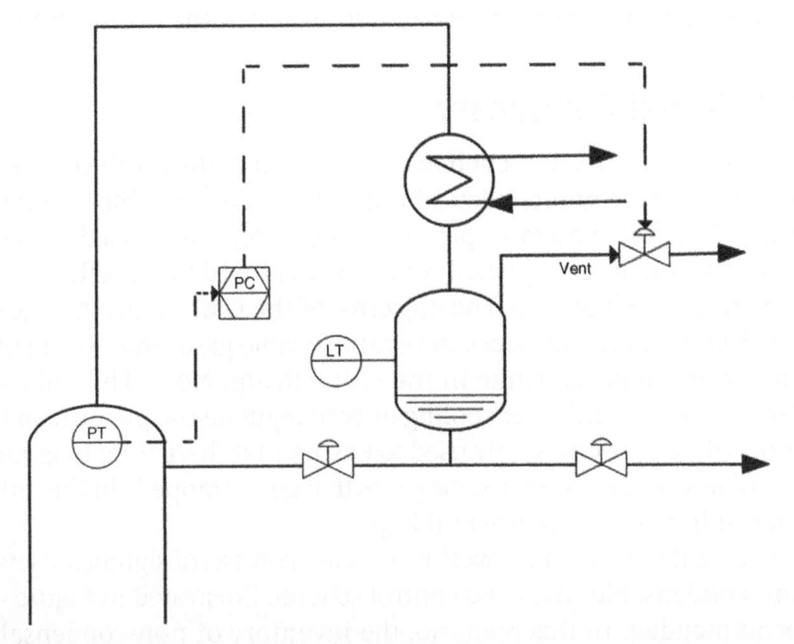

Figure 4.6 Pressure control with significant levels non-condensable gases.

4.2 Level Control

In distillation columns, there are at least two accumulators: one for the overhead vapor that is condensed and the other for the liquid leaving the bottom of the tower. The function of these accumulators, while seemingly simple, has generated much debate as to how the liquid inventory should be maintained. Thus, at this point in the discussion, we must segue to a topic that embroils many in unending debate: Level controller tuning. The debate revolves around whether accumulator levels should be controlled tightly, that is at set-point (SP), or allowed to "swing" even to the point of touching high and/or low-level alarm limits, as is often the desired objective when implementing "surge capacity control." In order to effectively address this topic, one must assess the role and purpose of accumulator vessels in the context of distillation tower service.

Given the importance of good inventory control, a short discussion on level control is warranted.

Taken alone, a drum or tank level (or a change to same) has no meaning or relevance: it is of little or no consequence where the liquid level is within the drum, as long as the drum is not emptied or overflows. That is, the vessel, and the fluid level within it, has no context. Once such a vessel is connected to a distillation column, however, as either a bottom's reboiler liquid sump, side-draw stripper/accumulator, or an overhead accumulator drum, then a *change* in level takes on a very significant meaning: it is the result of a change in mass or energy balance. While the objective of the bottoms and overhead accumulators is to help ensure proper inventory control, maintaining a consistent level in these two vessels has a direct influence on the tower's overall *mass and energy balance*. Thus, it is recommended that the level controllers on these vessels be tuned to respond to the level change *immediately* and return the level PV back to SP as quickly and stably as possible.

Given this insight on distillation tower behavior, one is compelled to wonder why the concept of "surge capacity control" has been so widely promoted? This question notwithstanding, the concept of utilizing surge capacity begs the question: What is/are the design criteria for distillation tower accumulators? The answer is that there are two, and ONLY TWO, factors affecting the size of distillation tower accumulators:

1. Protecting Rotating Equipment (e.g. pumps). That is, having enough liquid volume to ensure that a pump does not run dry and/or has sufficient NPSH, as well as enough vapor volume to allow a spare pump to be started should the primary pump stop running unexpectedly. In short, the drum volume is based on the material balance for the relevant streams such that the vessel does not overflow or go empty and to satisfy the specific rotating equipment requirements under nominal operating conditions.
2. Providing Residence Time for Phase Separation. This situation is found throughout refineries, as well as chemical plants, when immiscible components, usually water, need to be removed from hydrocarbon. The resulting vessel is significantly larger, by 2–4 times or more, compared with the first case, as the key design factor is residence time: allowing the fluid sufficient time to settle and be "undisturbed" such that the heavier component (again, usually water) can dissociate from the main component and be drawn away without entraining any of the main hydrocarbon liquid component with it.

So, from the above, it is apparent that "surge capacity," as it has often been applied for level controllers, is not a design criterion for distillation tower accumulators.

4.2.1 Surge Capacity Control

Based on the above discussion, one is likely to take the position that surge control in distillation columns should never be considered an option. Sadly, this is not always the case. On the (hopefully) rare case where surge control is warranted, the prudent engineer must understand:

1. Why using surge capacity is required, and
2. How the controls must behave in order to implement it properly.

To begin, the purpose of surge control is to prevent "large" disturbances from perturbing an entire operating unit by having one vessel absorb all – or at least a good portion – of the disturbance. The logical choice of vessel is a tank, drum, or accumulator whose sole purpose for existence is to provide some degree of fluid hold-up within the process. That is, to provide a buffer within the process to absorb disturbances.

The second point, as to how surge controls should behave, requires one to understand that there are two (2) opposing objectives at play for properly implementing surge control:

a. Maintaining the fluid inventory within the vessel (i.e. do not overflow or empty it);
b. Minimize the change in flow to the downstream equipment.

Given this dichotomy, the ability to satisfy both objectives depend on the magnitude of the disturbance, relevant to the vessel's available surge capacity. If the disturbance is "large" (relative to the available volume), then minimizing the impact to the downstream equipment may well be physically impossible. Hence, the challenge in implementing surge control using vessels that are not designed for that purpose!

One other point that is worth mentioning: There are two (2) different scenarios that will affect the difficulty of implementing surge control. That is whether the unit has a "once-through" configuration (such as a typical multicomponent separation train) or it has a circulation or recycle configuration (such as with extractive solvent distillation or an amine absorbing unit) wherein the process equipment and associated elements (flows, levels, pressures, etc.) interact with each other. The former is the easier one to implement as the "once-through" process, by its nature, lends itself to isolating the disturbance. Whereas the recycling process causes disturbances to circulate throughout the entire process, resulting in either amplifying its effect as the disturbance propagates and recycles throughout the unit or taking a "long" time to dampen out the effect (e.g. oscillations).

In both cases, when assessing how to implement surge control, the controller behavior should make output changes that:

A. STOPs the level movement with the smallest possible output movements;
B. Prevent the inventory (level) from exceeding upper and/or lower operating and/or alarm limits (e.g. 80% and 20%);
C. Once A&B are satisfied, slowly returning the level to the operating target (i.e. SP).

For both cases, the critical choice is determining *WHERE* the disturbance will be absorbed (if it is even possible to do so; in many cases, it would not be).

The obvious first choice is the vessel that is closest to the disturbance source. If that vessel is insufficient, then there are two (2) alternatives to consider:

i. Propagate the disturbance downstream to another, more suitably-sized vessel, or
ii. Distribute the disturbance across multiple vessels (the most appropriate choice for the recycling process and more challenging to achieve).

Of course, a fair amount of experimentation is often required to find the right balance, preferably with a dynamic simulation (though a low fidelity simulation is perfectly fine) or the actual plant. Based on the formulae provide in Section 4.2.3, however, a crude, but reasonable, simulation could be built using a spreadsheet or a custom program with which to experiment. The only additional information required for the simulation is the PID formula used by the control system (which ensures that the tuning that is developed in the simulation is directly transferable to the plant control system).

The point here is that, while surge capacity control has its place, it is the exception to the rule in the context of distillation. Therefore, one should clearly document the purpose and reason(s) why it is implemented. Furthermore, where it is found to be necessary, then the mechanical design of the affected equipment should be assessed to determine how it may be altered such that the surge controls are not required, which, may then, require replacing one or more vessels.

So, as described above, it is apparent that "surge capacity," as it has often been applied for level controllers, is NOT a design criterion for distillation tower accumulators. For further clarity to this topic, the only time one should tune a level controller to utilize surge capacity is when the drum is specifically designed to act as a surge vessel. That is a tank specifically and intentionally located between two units to absorb flow variations between them. Furthermore, in terms of distillation tower design principles, the bottoms and the overheads *accumulators do not exist*; that is, there is no accounting for any liquid holdup or inventory in either a steady-state simulation or any of the other process design methodologies that are currently employed (as mentioned in Chapter 2). Consequently, these vessels are effectively a "wide spot" in the pipe and the controls must treat them as such.

Lastly, any "sloppiness" in level controller response, from implementing either "surge capacity control" or other tuning and/or controller configuration that lengthens the time required to correct for a mass or energy balance disturbance, artificially lengthens and distorts the true dynamic characteristics of the tower's quality measurements versus MVs used to affect them, resulting in less robust quality controls, which ultimately results in greater quality variability. Therefore, based on all the points enumerated above, distillation column accumulator levels MUST be controlled tightly!

To be sure, there may be situations where downstream equipment constraints limit how much and how fast a particular-level controller MV can be moved to maintain a vessel's inventory. However, one must recognize that *every* unmitigated disturbance, no matter how small, **will** impact the tower's key quality variables, thus adding to the overall process variability, which, in turn, limits how closely the tower can be operated to its economic optimum. In short: **there is a penalty to be paid for loose or sloppy control**. So *be very selective* about where and how surge capacity controls are implemented on distillation towers. This caveat notwithstanding, it is recommended to avoid "surge capacity control" or any other form of "loose" control on distillation towers, if at all possible.

4.2.2 Open-Loop Stable versus Integrating Processes

The process design of an accumulator system can take one of two different forms, as illustrated in Figure 4.7. The distinction between the design lies in how liquid is drained from the vessel: it can be either gravity based or pump assisted as shown below.

While Chapter 7 focuses explicitly on the controller tuning, it is important to touch upon two distinct type of process responses that are expected accumulator vessels. Respectively, these two types of behaviors are referred to as an *Open-Loop Stable* (OLS) *Process* and an *Integrating Process*. The *Integrating Process* is a cumulative (or integrating) function of mass (im)balance over time. The difference between the two responses is shown in Figure 4.8, which illustrates an open-loop step test for both types of processes.

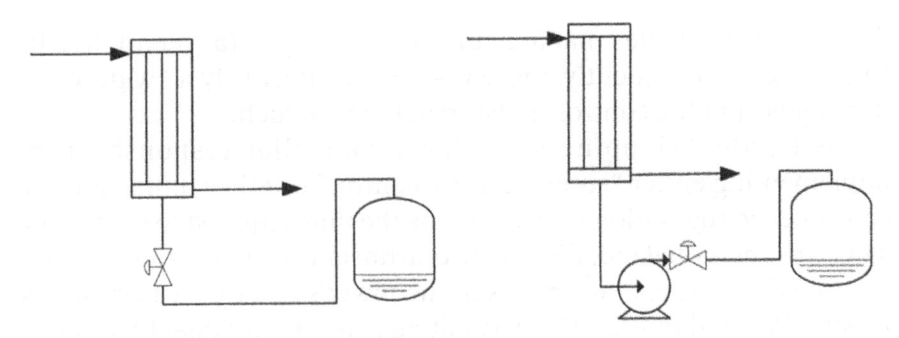

Figure 4.7 Different designs of accumulator systems.

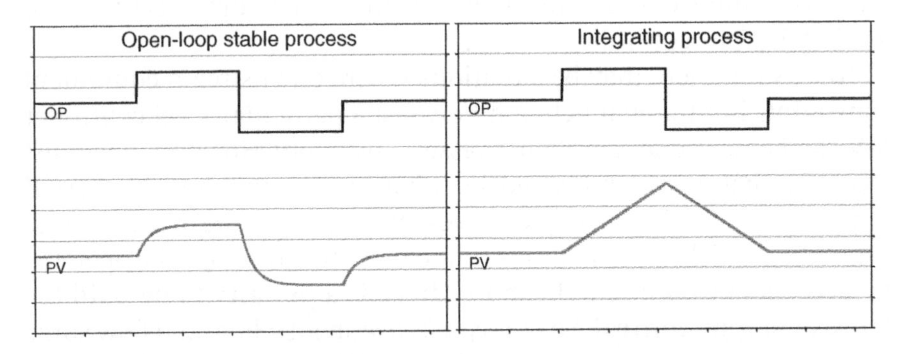

Figure 4.8 Open-loop stable versus integrating process step test response.

As illustrated in Figure 4.8, the behaviors are very different and requires significantly different controller tuning to achieve robust and stable operations. Refer to Chapter 7 for further details.

4.2.3 Calculating the *Process Gain* for Vessel Levels

Setting aside the debate over level controller tuning, one of the very convenient aspects regarding vessel level control is that, unlike many other controls (that would otherwise require a high-fidelity dynamic process simulator), the Process Gain (K_p), as seen by the level controller, can be calculated a priori – *before the equipment is installed, much less operational!* This is where performing a RGA is easiest and pays great dividends for both new plant designs and existing units. The derivation for calculating K_p for levels follows.

The mathematical representation for a change in level is derived from the mass balance equation where steady state is not assumed, as shown in Eq. (4.2).

$$\frac{\Delta \text{Volume}}{\Delta t} = \left(\Delta F_{Output} - \ F_{Input}\right) \tag{4.2}$$

where, for the purposes of controller tuning or dynamic modeling, Δt is the controller execution period or the simulation time step; the units for flow and volume are also consistent with each other, as well as with the time units.

4.2.3.1 Vertical Cylinder Vessels

For a vessel configured as a vertical cylinder, where the cross-sectional area remains constant over the span of the level meter, the change in volume is calculated as follows:

$$\Delta \text{Volume} = \text{Area} \times \Delta L \tag{4.3}$$

Substituting Eq. (4.3) into Eq. (4.2) yields:

$$\frac{\text{Area} \times \Delta L}{\Delta t} = \left(\Delta F_{Output} - \Delta F_{Input}\right) \tag{4.4}$$

The *process gain*, as seen by the controller, is determined by substituting the change in level (ΔL) with the expression for a change in the controller's *process variable* (PV), as given in Eq. (4.5).

$$\Delta L = \Delta PV \times \frac{\text{LM}_{Span}}{100\%} \tag{4.5}$$

where ΔPV is the change in level measurement seen by the controller and LM_{Span} is the calibrated range of the level meter. Level meters, especially those based on delta-pressure measurement, are often calibrated in "inches H_2O" but must eventually be converted to a physical displacement equivalent to the units of the vessel dimensions. Transmitter ranges are used because many control systems do not use *Engineering Units*, but instead, perform control calculations based on *Percent-of-Scale* or *Range-Based* values. Many control systems also utilize range-based values for all internal calculations. The advantage of this is that the hardware only need support integer math processing rather than floating point processing. This harkens back to the time when hardware costs, especially for floating point processing, were quite high. Although this makes computing more efficient, it puts the onus on engineers and technicians to update Controller Gain (K_c) values whenever instrument ranges, for either the controller PV and/or a secondary controller's PV (if applicable), or the control valve's maximum flow are changed. Therefore, all the derivations that follow will generate *Range-Based* or Percent-of-Scale values.

Substituting Eq. (4.5) into Eq. (4.4) yields:

$$\Delta \text{PV} \times \frac{\text{Area} \times \text{LM}_{Span}}{\Delta t \times 100\%} = \Delta F_{Output} - \Delta F_{Input} \tag{4.6}$$

If the level controller manipulates an output flow controller, then an expression relating the controller Output to that flow is required, as shown in Eq. (4.7).

$$\Delta F_{Output} = \frac{\text{FM}_{Span}}{100\%} \times \Delta \text{OP}_{Out} \tag{4.7}$$

where OP_{Out} is the level controller output.

If the controller manipulates an In Flow, then the sign on the calculated gain is negative. In either case, however, it is the numerical value of the process gain that is of interest, as most control systems rely on a "Control Action" parameter to address the sign of the process gain.

Substituting Eq. (4.7) into Eq. (4.6) and disregarding the input flow (F_{Input}), since for this purpose, it is an unmeasured disturbance, which yields the formula in Eq. (4.8).

$$\frac{\Delta \text{PV}}{\Delta \text{OP}_{Out}} = \frac{\Delta t \times \text{FM}_{Span}}{\text{Area} \times \text{LM}_{Span}} = K_{p_{Range-Based}} \tag{4.8}$$

Substituting the Area term with the formula for the area of a circle, the *Range-Based Process Gain* is calculated, as shown in Eq. (4.9).

$$K_{p_{Range-Based}} = \frac{4 \times \text{FM}_{Span} \times \Delta t}{\text{Diameter}^2 \times \pi \times \text{LM}_{Span} \times \text{SG}} \tag{4.9}$$

where SG is the specific gravity of the process fluid (if the flow units are mass-based, otherwise it is not needed).

While the vast majority of liquid-containing vessels are cylindrical, there are a few that are rectangular, such as cooling tower basins. Substituting an equivalent diameter (relevant to the enclosed area) or replacing the circular area formula with an appropriate value or formula also works!

If the level controller directly manipulates a control valve, then FM_{Span} is substituted with the control valve's maximum flow. Appropriate unit conversion factors are also required to cancel out the engineering units and obtain the result in *Percent-of-Scale*.

The inherent assumption is that the valve will always pass the maximum flow under all process conditions. However, hydraulic disturbances may prevent this from happening or otherwise alter the maximum flow the valve can pass. Where such disturbances are frequent and significant, additional measures are required to achieve stable, reliable, and robust control. The details for how to account for such disturbances is beyond the scope of this book, but a little careful thought should reveal the solution!

The formula in Eq. (4.9) assumes there are no internal baffles in the vessel that reduce the total volume of fluid the level controller is measuring. If there are baffles, then the Area term should be adjusted appropriately; refer to the section on Horizontal Drums for guidance with cylindrical vessels.

Now that the Process Gain is known, tuning rules can be applied to provide the desired controller response for assumed process dynamic characteristics. It is also recommended that, even if the control system uses engineering unit-based controller gain values, the range-based process gain should still be determined as inverting this value provides a validation of an appropriate MV–CV pairing that will generate acceptable operational behavior. A relative gain analysis (RGA) should be performed using the formulae provided herein.

For most level control problems, the only contribution to *process dynamics* comes from the control valve's response to changes in air load: the dynamic contribution is Lag Time only, no Dead Time. This is assuming that the level controller manipulates either a control valve or a vessel's liquid in/out flow controller. If the level controller manipulates, say, a reboiler duty, then more complex dynamic behavior should be expected (i.e. Dead Time + First Order Lag). While there may be, in fact, higher order dynamic effects, a FO + DT model is all that is required (and all that should be used) for a PID controller. Typically, this Lag Time is very small (usually much less than 30 seconds), so level controllers can be tuned aggressively.

Where more complex dynamics exist, for example with a side draw flow that is used to control a bottoms level or boiler feed water controlling a steam drum level, then a First Order Plus Dead Time (FO + DT) model must be obtained from a step test to determine the dynamic characteristics of the MV–CV relationship.

4.2.3.2 Horizontal Cylinder Vessels

Unlike a distillation column bottoms accumulator that has a constant cross-sectional area over the span of the level meter, the relationship between the level indication and volume for a vessel configured as a horizontal cylinder is variable and nonlinear. This relationship is defined by Eq. (4.10).

$$\Delta\text{Volume} = L \times \Delta A \tag{4.10}$$

where L is the tangent-line-to-tangent-line (T-T) *Length* of the vessel. The ΔA term is then given in Eq. (4.11).

$$\Delta A = \text{Area} \times \frac{\Delta\%\text{A}(\text{PV})}{100\%} = \frac{\text{Dia}^2}{4 \times 100\%} \times \Delta\%\text{A}(\text{PV}) \tag{4.11}$$

where $\Delta\%A$ (PV) is the change in Percent Area as a function of the level indication (i.e. PV). Substituting Eqs. (4.10) and (4.11) into Eq. (4.2) and rearranging yields:

$$\Delta\%A(\text{PV}) = \frac{4 \times \Delta t \times 100\%}{L \times \text{Dia}^2} \times \left(\Delta F_{Output} - \Delta F_{Input}\right) \tag{4.12}$$

To determine the *Process Gain*, the equation must be in the form of $\Delta\text{PV} = f(x)$. This transformation is made by dividing Eq. (4.12) by the *derivative* of the relationship between the Area and Level, as shown in Eq. (4.13).

$$\Delta\text{PV} = \left(\frac{\Delta\text{PV}}{\Delta\%A(\text{PV})}\right) \times \frac{4 \times \Delta t \times 100\%}{L \times \text{Dia}^2} \times \left(\Delta F_{Output} - \Delta F_{Input}\right) \tag{4.13}$$

Disregarding the input flow (ΔF_{Input}) and substituting the output flow with its range-based equation yield the equation for the range-based *Process Gain*, shown in Eq. (4.14), where $f'(\text{PV})$ is the *derivative* of the Percent Area as a function of level.

$$K_{P_{RangeBased}} = \frac{4 \times \Delta t \times \text{FM}_{Span}}{\text{Diameter}^2 \times L \times \text{SG} \times f'(\text{PV})} \tag{4.14}$$

The astute reader will likely notice that Pi (π) is seemingly missing from the formula in Eq. (4.14). It is, in fact, not missing but is captured in the $f'(\text{PV})$ function. The proof of this assertion is left as an exercise for the reader!

To complete the formula in Eq. (4.14), a function that relates Percent Area to PV is needed. This function is obtained by first taking the area of a circle and subtracting the area enclosed by a chord that represents the fluid level and the circle perimeter above the level, as shown in Figure 4.9.

The area enclosed by the chord l and the upper perimeter of the circle (the Area of the "Segment") is given in Eq. (4.15), where the units of the angle θ is radians. The angle θ is determined from Eq. (4.16) as a function of the distance d and the radius of the circle. However, to relate this back to a vessel's liquid level, the argument of the trig function is replaced with Eq. (4.17):

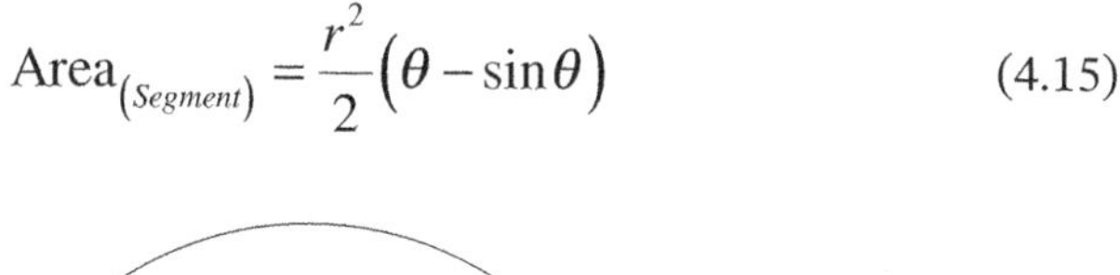

$$\text{Area}_{(Segment)} = \frac{r^2}{2}(\theta - \sin\theta) \tag{4.15}$$

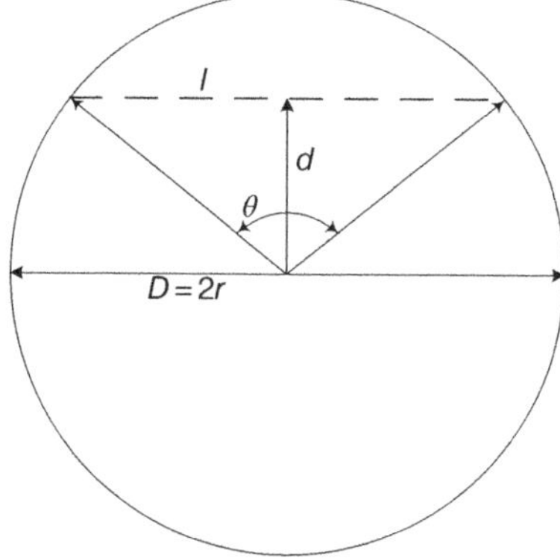

Figure 4.9 Truncated circle parameters.

$$\theta = 2\cos^{-1}\left(\frac{d}{r}\right) \tag{4.16}$$

$$\frac{d}{r} = \frac{\text{PV} - 50}{50} \tag{4.17}$$

Substituting these into Eq. (4.15), subtracting the result from the Area of a circle and then dividing that result by the Area of a Circle gives an equation for calculating the Percent Area of a Truncated Circle (Circle minus a Segment) as shown in Eq. (4.18):

$$\%\text{Area}_{(Circlet)} = 100 - \frac{50}{\pi}\left[2\cos^{-1}\left(\frac{\text{PV}-50}{50}\right) - \sin\left(2\cos^{-1}\left(\frac{\text{PV}-50}{50}\right)\right)\right] \tag{4.18}$$

The results of this equation can be regressed to a third-order polynomial for the entire PV range (0–100%) as shown in Eq. (4.19) and Figure 4.10.

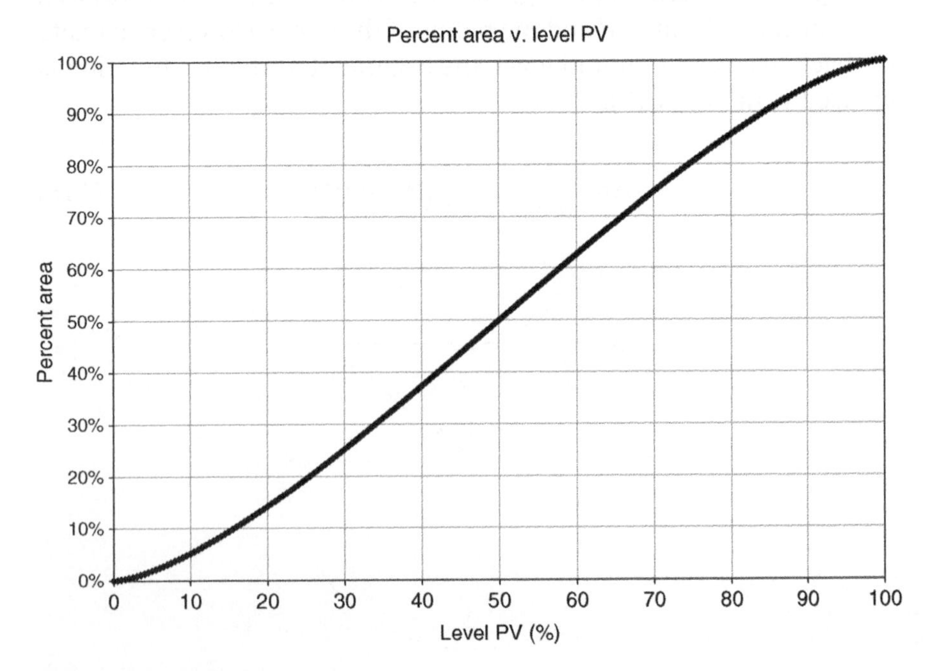

Figure 4.10 Graph of Eq. (4.18).

$$\%\text{Area}(\text{PV}) = -0.78047 + 4.6868\times10^{-1}\text{PV} + 1.6408\times10^{-2}\text{PV}^2 - 1.0939\times10^{-4}\text{PV}^3 \tag{4.19}$$

The derivative is easily obtained, as given in Eq. (4.20), and is substituted into Eq. (4.14) to calculate the *Process Gain* at a specified level; refer to Figure 4.11 for a plot of Eq. (4.20). *This assumes the level meter span is equal to the vessel diameter* and no correction is made for liquid volume contained in the heads. Refer to the chapter on controller tuning for an analysis of these effects and the potential impact on the observed *Process Gain* by a level controller.

$$f'(\text{PV}) = \frac{d(\%\text{Area})}{d\text{PV}} = 4.6522\times10^{-1} + 3.2977\times10^{-2}\text{PV}^2 - 3.2977\times10^{-4}\text{PV}^2 \tag{4.20}$$

To compensate for the difference between the total vessel volume and the portion measured by the level meter, the area term in the *Process Gain* formula must be adjusted to account for the liquid volume above and below the level meter's nozzles. Eq. (4.21) is the formula

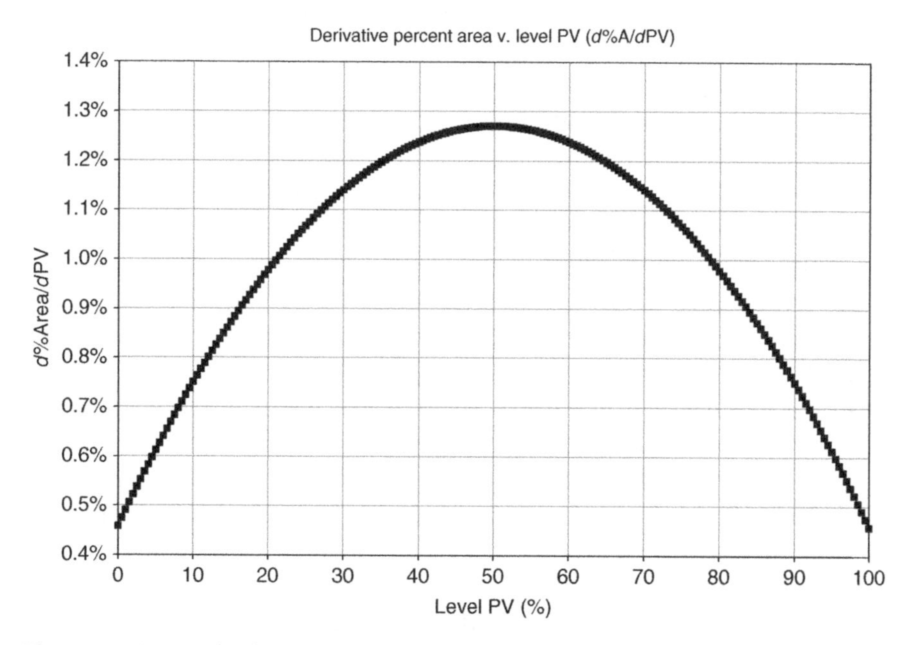

Figure 4.11 Graph of Eq. (4.20) (Eq. (4.19) derivative).

that provides the fraction of the vessel's volume that is measured by the level meter. This formula amounts to a "correction factor" that is applied to the *Process Gain* calculation.

$$\text{Area}_{adj} = \left[\%\text{Area}\left(\frac{\text{LM}_{Offset} + \text{LM}_{Span}}{\text{Dia}}\right) - \%\text{Area}\left(\frac{\text{LM}_{Offset}}{\text{Dia}}\right)\right] \Big/ 100 \quad (4.21)$$

where LM_{Offset} is the distance between the bottom of the vessel and the lower tap of the level meter, LM_{Span} is the level meter's calibrated span and %Area(x) is the formula in Eq. (4.18). This correction factor is applied in the denominator of Eq. (4.14).

4.2.4 Relative Gain Analysis, aka Closing the Loop in Plant Design

The base design parameters for every control element, pump, compressor, and vessel within a process unit is obtained from a steady-state model of the process. Base regulatory controls are added to the design at a relatively early stage. Detailed design of the vessels, heat exchangers, and pumps follow with some recycle occurring on the sizing of piping and instrumentation to account for the equipment details. The ability of the regulatory controls to achieve their objectives in the context of the actual equipment design, however, is rarely revisited. This has led to undesirable results when the unit is put into service: namely certain rudimentary controls are incapable of affecting the process as intended, resulting in constrained, subpar (below design) plant performance, particularly around distillation trains. This situation is easily avoided by applying some fundamental principles and performing some very simple analysis as the detailed equipment design nears completion.

As indicated previously, a RGA comparing alternative MV–CV pairings, particularly for inventory controls, can be extremely useful to determine the best pairing. While typical overhead and bottoms accumulators for binary distillation columns (e.g. those with only a distillate and bottoms product streams) have limited options, the RGA can still provide important insights.

The insights amount to whether (or how well) any of the available MVs can maintain the liquid inventories. The tower shown in Figure 4.12 is a typical binary distillation column; as described in Chapter 2, there are four (4) MVs available for inventory control:

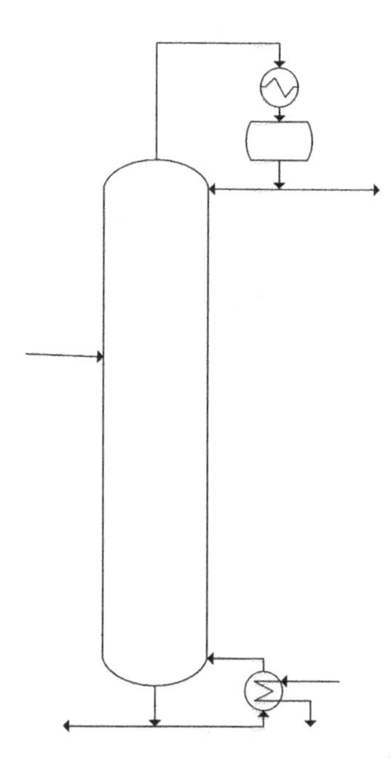

Figure 4.12 Typical binary distillation column.

a. Distillate Flow
b. Bottoms Flow
c. Reboiler Duty
d. Reflux Flow

Discounting, for the moment, the interdependence between Reflux Flow and Reboiler Duty, the RGA should assess the impact of both of these streams on their respective levels: the Overhead and Bottoms Accumulators, respectively (recalling that prospective MVs should be selected based on their proximity to the CV of interest; refer to Section 2.3). The reason for performing this analysis is illustrated in Figure 4.13, wherein the level controller response for the MV–CV pairings with decreasing K_p are provided – for both an SP change and an unmeasured disturbance. As illustrated, the level controller must respond with increasing magnitude changes in the MV as the K_p gets smaller in order to achieve the same level behavior.

The formulae given in Eqs. (4.9) and (4.14) provide the basis for assessing the flows out of the accumulators. But what about the effect of Reboiler Duty? This analysis requires including the heat of vaporization and the process fluid and either the sensible heat or heat of condensation of the reboiler duty to the equivalent mass of process

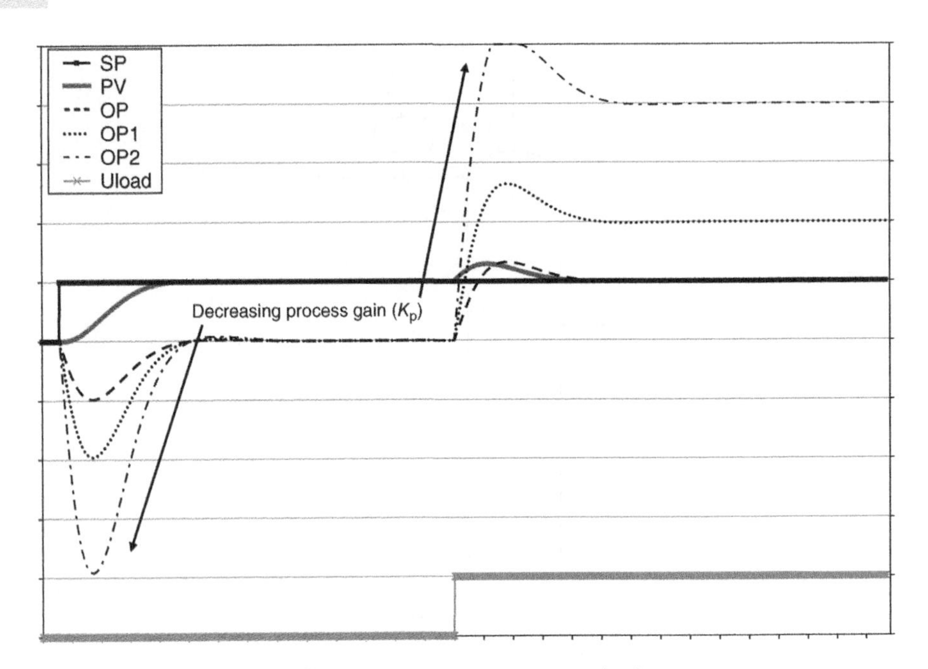

Figure 4.13 Level controller response as process gain decreases.

fluid. While an explicit accounting of heat transfer efficiency may be excluded for steam or hot oil reboilers, when using a fired heater, it may be warranted.

While the above analysis may appear to be a bit overkill for the very simple distillation tower example, as the number of degrees of freedom increase (say, from the addition of one, or more, side draws), the relevance of the analysis should become apparent. The decision on appropriate MV–CV pairing requires such an analysis to ensure the running process achieves the design intent. The RGA will reveal which stream(s) have the greatest influence over inventory controls and, thus, which MV–CV pairing(s) should be considered. While large disparities between MVs for a specific CV make the final choice clear, as the K_p's begin to approach parity, that is when other effects must be considered: namely the effect on (product) stream composition. This is covered in Chapter 5.

This concludes the chapter on distillation inventory controls. In addition to reading on the reader is referred to the following references for some of the ideas developed herein – Shinskey (1977), Sloley (2001), and Svrcek et al. (2014).

Tutorial and Self-Study Questions

4.1 What are the characteristics of common flow loops?
4.2 Describe the characteristics of common liquid pressure loops?
4.3 Describe the characteristics of common liquid level control loops?
4.4 Describe the characteristics of common gas pressure control loops?
4.5 Why should pressure and level controls in a distillation tower be established prior to attempting quality control?
4.6 Imagine a situation where you have a total condenser, and you are using the condenser duty (cooling water flow) to control the pressure. However, over time you are gradually losing controllability of the system. Based on your distillation control understanding explain why this maybe happening and how to control it? What additional components needs to be added? How would you control this situation?
4.7 Why is it important to understand the geometry of the condenser and reboiler vessels?
4.8 What makes (most) levels (and some pressures) different from other process measurements used for control? How does this affect tuning? In what situation does a level behave the same as a flow?
4.9 What was (is) the purpose of a hot-vapor bypass (HVB) on a distillation tower overhead pressure control design? Why was it useful? Is it still required? When should it NOT be used?
4.10 What are the most important physical (mechanical, piping, layout, etc.) design considerations for flooded condensers (that ensure good control)?

References

Shinskey, F.G. (1977). *Distillation Control for Productivity and Energy Conservation*. McGarw-Hill.
Sloley, A.W. (2001). Effectively control column pressure. *Chemical Engineering Progress* Mar: 75–83.
Svrcek, W.Y., Mahoney, D.P., and Young, B.R. (2014). *A Real-Time Approach to Process Control*. Wiley.

5 Distillation Composition Control

Once inventory control on a distillation column is established, then the focus shifts to achieving the columns operational objectives: making on-specification products. As discussed in Chapter 2, this entails the manipulation of the mass and energy balances in order to achieve the desired product specification. To this end, distillation composition control, in any form, requires the determination of the component content at the product stream. While the "product" is, generally, the distillate stream, it may often be the bottoms stream, as is often the case in refining units (e.g. demethanizer, depropanizer, debutanizer). Ideally, this determination should be made by measuring the composition directly, for example, using a gas chromatograph or other analytic instrument. However, these types of measurements carry significant installation and operational cost, can exhibit large time delays, and they require dedicated maintenance to ensure robustness. As a result, in general distillation designs composition control is achieved by using temperature as a surrogate or inference for composition.

5.1 Temperature Control

Temperature sensors are inexpensive, highly reliable, repeatable, continuous, and fast compared to composition sensors (e.g. Fruehauf and Mahoney 1994; Svrcek et al. 2014). The measurement lag is particularly important for dynamic considerations. For temperature it is a fraction of a minute, whereas composition measurement by gas chromatography

A Real-time Approach to Distillation Process Control, First edition. Brent R. Young, Michael A. Taube, and Isuru A. Udugama.

Companion website: www.wiley.com/go/Young/DistillationProcessControl

is on the order of 5–10 minutes for typical impurity measurements of >1% but may be as long as 20 minutes, or more, for high-purity measurements in the PPM range. Infrared or other photometric devices that produce continuous composition estimates are seeing increased use and response times are of the order of a minute (e.g. Tanaka et al. 2010). In either case, periodic checks of product composition by analytical means provide information that is used to update the temperature control target. The accuracy of the correlation between column temperature and product composition depends on the sensitivity of the controlled temperature to composition changes and pressure variations at the temperature measuring point.

The sensitivity of the temperature measurement to key or major component compositional changes for each tray can be determined if tray-by-tray composition changes are large and the other component changes are small. It must be determined which stage exhibits composition-related temperature response in all disturbance situations. A sizeable temperature response must be present for all process variable changes to which the column will be subjected. Select a range of process disturbances and change these in short step sizes to compare the tray temperature profiles (e.g. Trevedi 1993). A temperature measurement in this area will give a good indication of composition.

The temperature composition correlations of key components are often affected by changes in the concentration of other components (i.e. column feed composition changes). If the magnitude of these changes can be estimated, a calculation using equilibrium constants can be made to determine the effect on the temperature composition correlation. Then a control tray can be selected where the effect of non-key component variations is small.

Stable column temperature control, from the tray selected by the foregoing static considerations, depends on the dynamic measuring lag or response of the tray temperature with respect to the manipulated variable used to control the temperature. Based on experimental tests, the following observations are cited for use as guides:

1. Temperature control is made less stable by thermowell and measuring instrument lag or response times.
2. The speed of response and control stability of tray temperature, when controlled by reboiler heat, is the same for all tray locations.

3. The speed of response and control stability of tray temperature, when controlled by reflux, decreases in direct relation with the number of trays below the reflux tray.
4. When pressure is controlled at the temperature control tray, the temperature instrument's response speed can vary considerably with tray location and is normally slower.

5.1.1 Setting Up a Single Temperature-Based Composition Controller

To understand the setup process of a single temperature-based composition controller, let's look at the generic distillation column configuration in Figure 5.1.

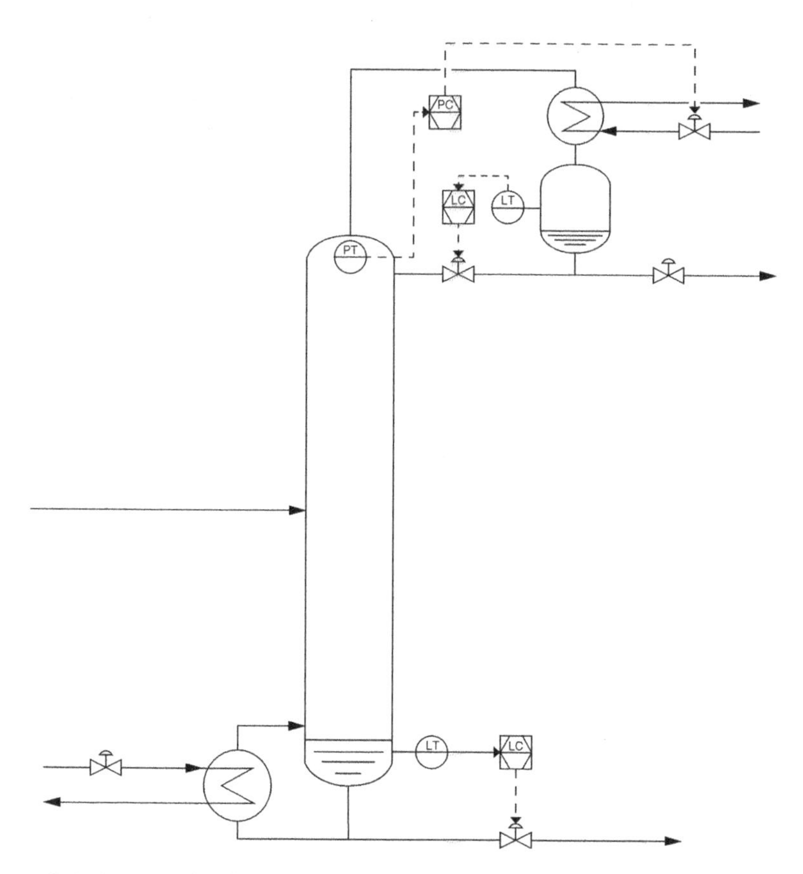

Figure 5.1 Generic distillation column configuration.

The generic distillation column configuration illustrated in Figure 5.1 has three control loops that have been designed, namely they are

1. Pressure control loop (manipulating condenser duty)
2. Reboiler level (manipulating bottoms flow rate)
3. Reflux drum level (manipulating reflux flow rate)

Let's assume that these controllers have been designed based on a gain analysis as described in Chapter 2 and the guidelines set out in Chapter 4. So, if the composition at the distillate stream needs to be controlled, the only viable control loop that can be manipulated would be the distillate flow rate, resulting in the single temperature-based composition controller illustrated in Figure 5.2. Note: the temperature measurement will be taken on a tray based on the criteria set out in the previous section.

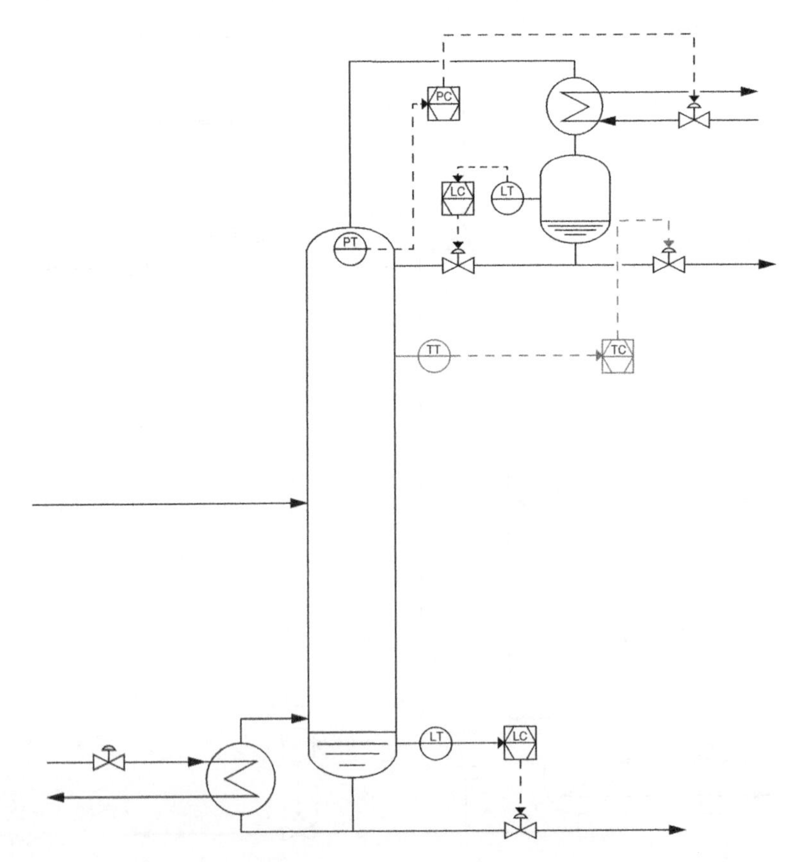

Figure 5.2 Single temperature-based composition control for the distillate.

5.1.2 When Temperature Is Like an Integrating Process

This section heading may seem like an oxymoron. Still, the point is that there are, in fact, situations where a distillation column (or other process equipment) temperature behaves exactly like a tank level, that is as an *integrating process*. While not rare, these situations are usually infrequent. Nevertheless, the frequency of their occurrence seems to be increasing as specification limits become more stringent for refineries, LNG/LPG, and other plants.

There are two observed situations where a temperature behaves like an *integrating process*:

1. The first is when the component concentration (correlated to the temperature) is in the single digit with respect to mole percent.
2. The second is when water (or other cooling medium) is used to cool a process stream temperature using a hairpin, concentric pipe, or similar style heat exchanger.

In both cases, using tuning rules for an *open-loop stable (OLS) process* will produce undesirable behavior: the temperature will forever oscillate and only by coincidence it will find a "happy place" where the temperature lines out to some steady-state value. That is, when typical PID tuning (as might be used for an *OLS process*) results in the temperature continually oscillating, this oscillatory behavior is a good indicator of a temperature that behaves like an *integrating process*. Be aware that the "oscillations" may be "slow," that is with an oscillation period of several tens of minutes or, even, hours, depending on how the controller is tuned. The point is that it WILL oscillate because of the unrecognized underlying process behavior.

Therefore, the prudent engineer must be vigilant to look for and recognize such behavior and, subsequently, be prepared to address it with proper controller configuration and tuning. Keep in mind, however, that this may also be an indication of the presence of process Dead Time. In both cases, a step test should be performed to confirm the process characteristics of the CV as seen by the controller. Once the actual process behavior is known, then appropriate tuning rules can be applied to generate the desired controller behavior and process performance.

5.1.3 Reboiler Outlet Temperature Controls

The point of any temperature controller is to maintain a desired energy level (enthalpy) for liquids or composition (and sometimes enthalpy) for vapors. Therefore, the temperature measurement instrument must be installed in a location that provides a true representation of that property. Designers in the past, as well as today, have what seems like an undying affection for reboiler outlet temperature controllers, irrespective of the reboiler mechanical design. Without going through the historical background of reboiler designs, let's simply assert that, over time, designers failed to account for different types of reboiler designs and simply assumed that "a reboiler is a reboiler," so the controls should be the same!

An example of this assertion is from a specialty chemical plant. Per the plant's original design, reboiler outlet temperature controls were configured on all of the thermosyphon reboilers. In general, once the plant had been operating for some time, all of these had been replaced with either tray temperature or other parameter (i.e. reflux/feed, duty/feed) controls because it was found to be a better operational representation of the tower's performance and, thus, a better CV.

The main factor to consider regarding reboiler temperatures is to understand what process condition or property the temperature measurement is actually reflecting. More specifically, is the temperature measurement in a pure phase – only liquid or vapor – or an actual, or potentially, mixed phase? If it is the former (i.e. a pure phase), then the temperature is a reliable indication of enthalpy and/or composition; if it is the latter, then it is not.

So, for a reboiler outlet that is, or may be, a mixed phase, such as with thermosyphon or fired heater reboilers, a temperature measurement in that location fails to provide the desired information. If, however, it is a kettle-type reboiler with the temperature probe in the reboiler's vapor space or vapor outlet line, then this measurement is a valid indication of the fluid's thermal and composition properties.

To illustrate this point, the plot in Figure 5.3 shows the temperature profile in a refinery LPG splitter, each trace in the plot is a snapshot of the temperature profile at different times. The item to note in the plot is that the reboiler outlet temperature is higher than the bottoms (liquid) stream temperature, which is contrary to distillation column behavior in that temperatures should increase – get hotter – as one

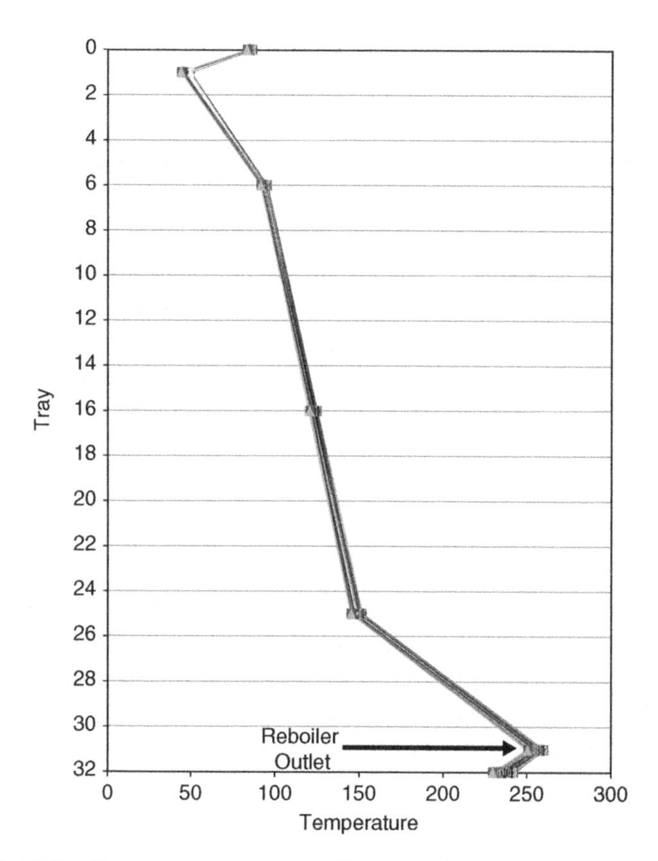

Figure 5.3 LPG splitter temperature profile.

goes from the top of the tower to the bottom. The explanation for this observation, however, is that this tower used a fired heater reboiler and when the reboiler effluent enters the tower it flashes; the energy absorbed by the process fluid in the heater goes toward evaporating some portion of the fluid, thus causing the liquid that remains to be cooler (compared to the mixed phase fluid temperature in the reboiler outlet line); the remaining liquid that accumulates in the bottoms accumulator is then in equilibrium with the vaporized fluid at this lower temperature.

The other "anomaly" that should be noticed is at the top of the profile that indicates the reflux temperature (Tray 1) is colder than the overheads vapor (Tray 0). This is due to the fact that the reflux liquid (where the temperature is measured) is subcooled (as described in Chapter 4).

Another important point to make here regards the use of the temperature profile in assessing an existing tower's behavior. This will be explored further in Section 5.5.

So, when assessing if or how a reboiler should be controlled, consider the fluid's physical properties as this determines whether or not a temperature measurement will be a meaningful indication suitable for control purposes. Furthermore, be aware that most process simulations do not, as a rule, provide an indication of the fluid state as it transits between the reboiler and the tower, but rather its state once inside the tower. So never simply rely on process simulations for "answers": A prudent engineer must THINK and understand what a simulation or calculation result is telling them.

5.2 Actual Composition Control

The composition control loops on a column are the most vital for operational success. The purpose of these controls is to maintain the product composition such that it satisfies the product quality specifications. This objective must be satisfied at all times, particularly in the face of disturbances and is often referred to as on-aim or minimum variance control.

While there are many factors that affect the dynamic behavior of MVs and disturbances upon a CV, such as transport delays in the process, how fast or slow a control element moves upon a change in signal, etc., it is only the final combination of these elements that truly matters: this is how the process *really behaves*! Thus, one must be prepared to deal with reality as found, at least for an existing plant. But even for the plant under design, there is only so much that can be done with or to the design that will affect the process' dynamic behavior.

With that in mind, recalling the instructions in Chapter 2 on MV–CV pairing, the MV of choice for composition control should be made such that the dynamic effects are "fast" (or as fast as possible) relative to the anticipated disturbances. This will ensure that product qualities are maintained – as well as can be – in the wake of expected disturbances: The Purpose of Process Control!

When the feed contains multiple components (as is usually the case) and their boiling points are relatively close together, fixing the temperature of a stage in the column may not fix the composition. Therefore, to help identify pertinent variables that could be used to elucidate the stage's composition under varying conditions, a steady-state model may be used to compare use of an online composition analyzer rather than a temperature controller. Factors to consider are yield loss, energy consumption, and dead time (e.g. Fruehauf and Mahoney 1993; Svrcek et al. 2014).

Sometimes the important disturbances can be measured or anticipated. In which case adding feedforward control is very good potential a candidate to consider. In other situations, the control loop structure can be rearranged to influence the way in which the controller responds to the disturbance effects on the composition variable. Several practitioners and researchers have proposed numerous algorithms for determining the disturbance sensitivity for different control structures. For example, Tyreus (1992) states that, in his opinion, direct dynamic simulation of the strategies resulting from assignment of the manipulated variables for pressure and level control gives the best insight into the viability of a proposed composition control scheme. Things to think about include:

- Practical considerations (questions to be considered are where to analyze and how often?)
- Consideration of a cascade (combines the benefits of faster temperature control with the potential accuracy of composition control). Figure 5.4 shows a composition temperature cascade control setup for distillation composition control.
- Energy and optimizing controls (composition control allows operators to balance energy usage versus throughput). After the inventory and composition controls have been assigned, there are typically a few manipulated variables remaining. These variables can be used for process optimization. Because process optimization should be performed on a plant-wide scale, in-depth discussion of this topic will be delayed until Chapter 10.

5.3 More Complex Control Configurations

The basic distillation control configurations have been described above. However, many other configurations use linear or even nonlinear combinations of the basic manipulated variables.

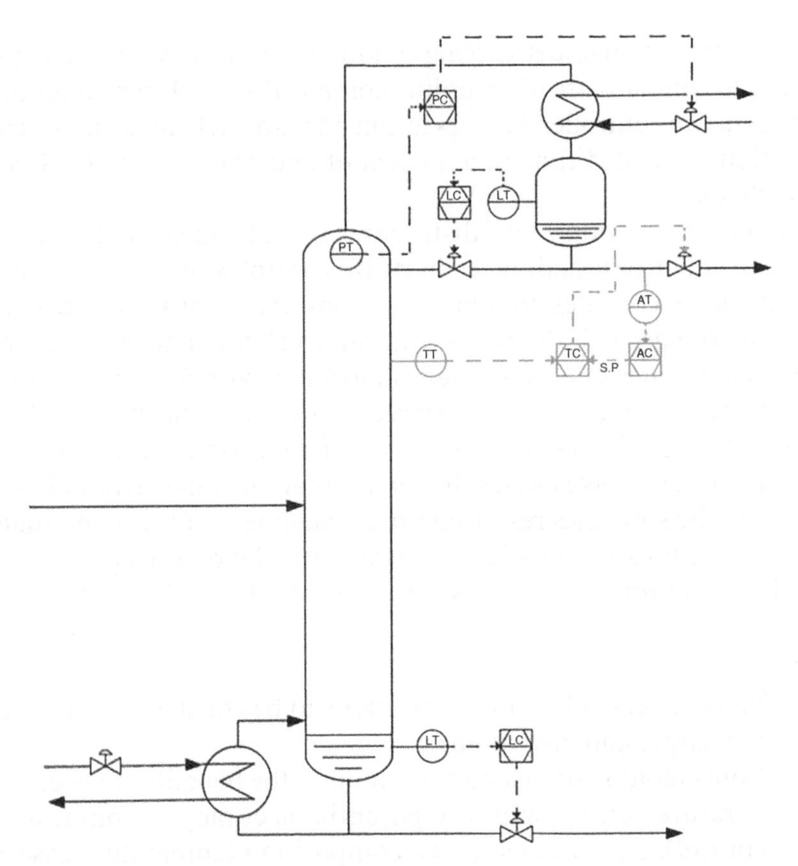

Figure 5.4 Composition to temperature cascade control for distillate composition control.

5.3.1 Ryskamp's Scheme

One common example, which is sometimes called Ryskamp's Scheme (1980), manipulates the reflux ratio (L/D), via ratio control, and the reboiler duty (V) as shown in Figure 5.5. While this scheme was implemented in years past, a careful analysis will reveal that proper level and temperature controller tuning eliminates the need for it.

Another relatively common scheme is the double ratio configuration, which manipulates the reflux ratio (as shown in the Figure 5.5 for Ryskamp's scheme) and also boil-up ratio similarly. This scheme has been widely recommended as it results in relatively small interactions between the two control loops. However, it often results in tower flooding as feed rates are increased, and it is the authors' view that it is

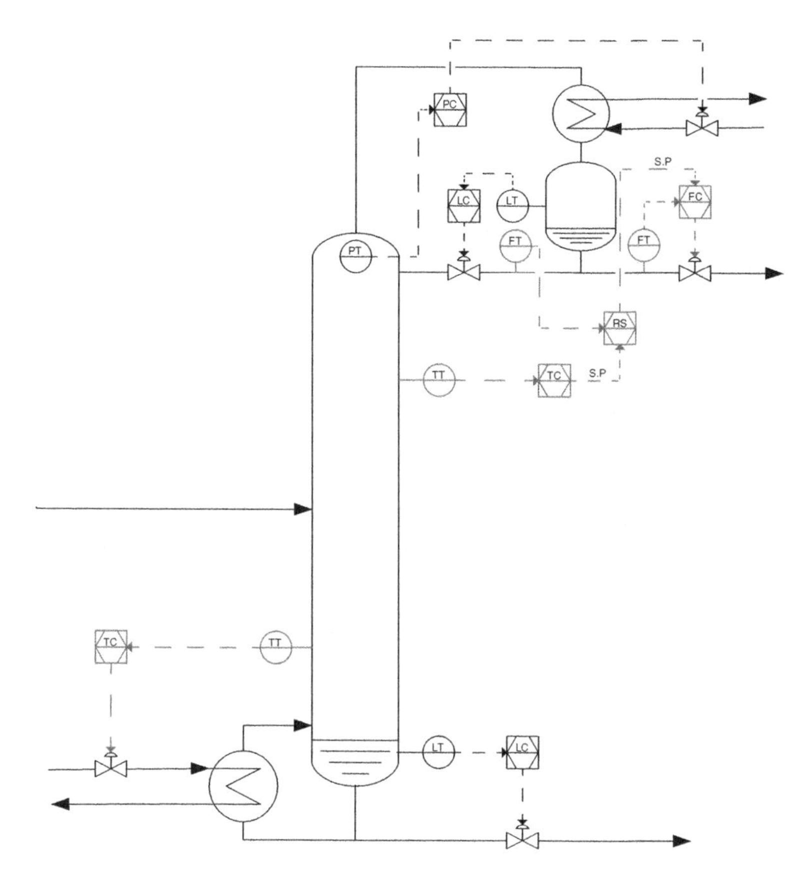

Figure 5.5 Ryskamp's scheme for distillate composition control.

overly complex – LV control configurations with adequate tuning will often achieve the same or indeed better results.

5.3.2 Dual Composition Control

Most industrial distillation columns operate similarly to the stabilizer we studied in the previous workshop. One degree of freedom is used to control a product composition and the second available degree of freedom is used to control fractionation or energy consumption. This mode of operation is often called one-point or single composition control.

Sometimes both the bottoms and the overheads products are equally important or equally valuable, and both the bottoms and the distillate compositions need to be controlled. This is called two-point or double-ended composition control and results in a much more difficult control problem than one-point composition control (Ryskamp 1980). The primary cause of the extra difficulty is the interaction that exists between composition loops in a distillation column. This property of distillation columns (inherent interactions between two or more control loops) is called ill-conditioning.

Among the problems created by ill-conditioning is that there usually exists only a very narrow operating range that satisfies both composition control loops. Essentially, the manipulated variables used for composition control need to be adjusted together to produce the required results. Many newer control schemes are based on this principle, including model-based control technologies, such as dynamic matrix control (DMC, the original MPC) (e.g. Cutler and Ramaker 1979; Prett and Gillette 1979).

A second problem with two-point composition control is that there are no degrees of freedom left to operate around equipment constraints such as a reboiler duty limitation or flooding limitation. One-point composition control schemes have one degree of freedom, which is not used for composition control and is available for this purpose (cf. a stabilizer that is operated at a fixed heat input). This can create operational difficulties for many industrial columns.

5.4 Distillation Control Scheme Design Using Steady-State Models

Steady-state simulation of distillation columns has become commonplace. The use of these simulations was originally restricted to determining heat and material balances and sizing purposes. Tolliver and McCune (1978) and others (e.g. Freuhauf and Mahoney 1992, 1994) have shown that steady-state calculations can be used to screen candidate control schemes, to provide a means for tray temperature location, and to calculate steady-state gains.

Steady-state simulation can often be used to evaluate options for base-level process control strategies early in the design, as described in Chapter 2. The advantage to this approach is that steady-state models are easily manipulated and the results are robust. This allows for the efficient generation of case studies necessary for steady-state design procedures. As the objective of this exercise is to assess the steady-state gains for different MV–CV pairings, the concepts of "control structures" doesn't really apply. Nevertheless, once the gains are determined for each MV–CV pairing, then the requisite control structures should be evident.

The basic steady-state design procedure consists of the following five steps:

1. *Develop a design basis*: Here there is a need to define product composition specifications, disturbance type and size, constraints, and original column design basis.
2. *Select a MV–CV pair*: While the literature abounds with multiple control configurations, it is usually best to start with a simple approach; if simple produces undesirable results, then more complicated structures can be tried.
3. *Conduct open loop testing*: The purpose of this step is to use the steady-state model to identify a suitable tray for the temperature sensor for composition control. The procedure consists of using a candidate MV, say, the distillate flow, and then observing the change in the column temperature profile. A good tray is one on which the temperature change was significant and nearly equal when the flow is both increased and decreased from the original value.
4. *Run closed loop testing*: In this step, the steady-state model is used to simulate the candidate MV and to test its robustness to feed flow and composition changes. This step consists of a series of runs (sensitivity studies) aimed at assessing how a set of operating conditions and changes to the subject MV affect the product specifications for the expected disturbances in feed flow and composition.
5. *Confirm the objectives have been met*: If the quality objectives have been met, then the procedure is complete: a suitable MV–CV pairing has been identified. If not, the procedure is repeated with another candidate MV.

As the simulation will ensure closure of the mass and energy balances, it is recommended to run the sensitivity analysis on only one MV at a time. Any material or energy balance effects to other streams should also be noted, as this often provides additional insights to how the inventory controls should be structured and the need for ratio or feed-forward controls.

To illustrate this point, consider, as an example, a typical 2-product or *binary* distillation column that has feed flowrate and composition as the disturbance sources. Only two degrees of freedom are available for such a column: feed split (mass balance) and fractionation (energy balance). The best feed split MV–CV pairings (i.e. those with the largest process gains) are direct, where one product stream is manipulated directly to control composition based on a tray temperature. The temperature controller manipulates the designated stream's flow: If the control temperature is above the feed to the tower, then the distillate stream is usually the MV of choice; the bottoms stream is the selected MV if the temperature is below the feed.

The case for using the bottoms flow is often due to the heavy components being a relatively small fraction of the feed. Thus, this stream may also be far too small to effectively control the bottoms accumulator inventory (which is often the default MV–CV pairing for the stream); an RGA of the tower's mechanical design will also reveal this.

It is also this second situation that leads to more *interesting* (i.e. less common) MV–CV pairings: The indirect feed split control is where the distillate flow is affected indirectly by manipulating the steam flow: The overheads composition is determined by a temperature controller that manipulates the steam flow.

The last part of the MV–CV pairing selection process is to determine if a ratio or feed-forward control is warranted, for example a reflux to feed flow or reboiler duty to feed ratio. Such structures will provide more robust control of the quality variables compared to simple feedback control structures.

5.5 Performance Analysis Using Steady-State Data for an Existing Distillation Tower

Using an offline steady-state process simulator has been shown to provide insights to the expected response of a distillation tower's critical quality measurements to changes in different MVs. This begs the question: (How) can steady-state data be used to assess an existing tower's behavior? More specifically, what insights can be gleaned about the most effective MV–CV pairing from historical plant data?

As alluded to in Section 5.1, a tower's temperature profile provides important information about what is actually happening inside the tower. Furthermore, combining the temperature data with analytic data (i.e. lab analysis and/or online analyzers) is another tool that should be in every process engineer's toolkit (Taube and Udugama 2021). A real-life example follows.

A "Lights Removal" tower was being assessed for poor performance and possible control structure changes. The first step in this analysis requires obtaining periodic snapshots of tray/stage temperatures and then creating a plot of tray/stage location versus temperature, as shown in Figure 5.6. While a single snap-shot in time is useful, having multiple snap-shots on the same plot provides insights to the variations in stage

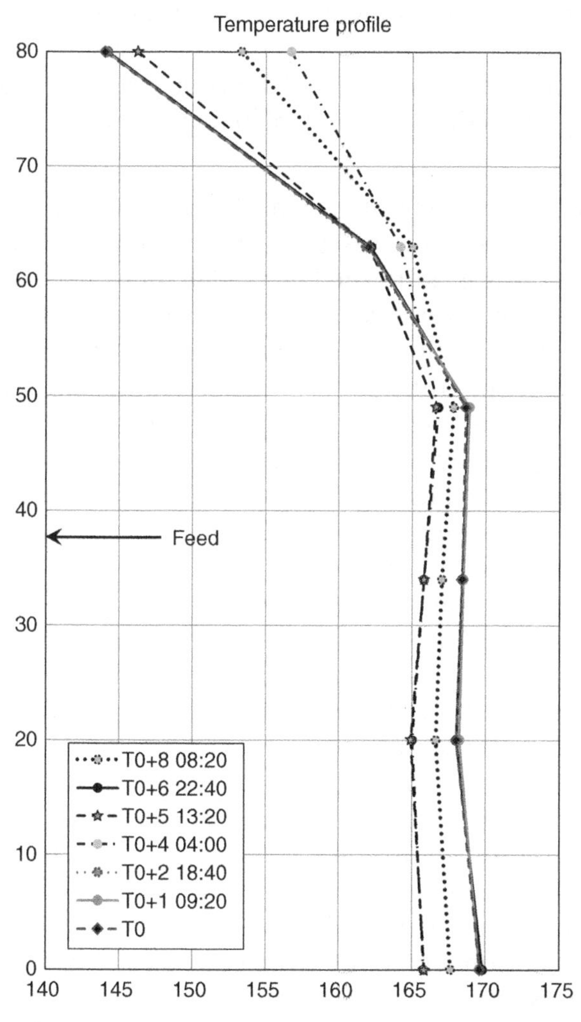

Figure 5.6 Distillation tower temperature profile plot.

temperatures, which may then be correlated to specific disturbance events (e.g. high or low feed rates, feed composition changes, diurnal or seasonal ambient temperature changes). The tray/stage temperature locations are based on either actual tray numbers or, in the case of a packed column, the distance of each temperature from some reference point: either the top or bottom of the tower. The objective here is to visualize the nominal temperature profile along the height of the tower and its variations over time. As seen in Figure 5.6, most of the separation is occurring above the tower feed, as the temperature profile below the feed is rather flat. Thus, as the tower description implies, the purpose of this tower is to remove the lights impurities from the product; the downstream tower then removes the heavy impurities.

The next step in the analysis is to correlate tray/stage temperatures with a stream composition analysis. Unlike the previous analysis, there are two (2) aspects to this analysis, which require special consideration:

1. Depending on the frequency of the composition analysis data, the timeframe of the assessment may be rather long. That is, if the analysis is performed by the lab only once to a few times per day, then several years of data is likely needed in order to capture the variations in operating conditions and its effect on the stream composition. If the analysis is available from an online device, such as a gas chromatograph (GC), then a shorter time frame of data may be used. But, this too, requires some judgment based on understanding of process history and how much variation occurs. Hence the multiple snap shots of the temperature profile analysis.
2. For lab-based analysis, there is often a very significant time delay between when the sample is taken from the process and when the analysis results are available. Thus, aligning the analytic data with the process data, which corresponds to the conditions when the sample was retrieved, is vital. While requiring some effort, this time alignment is accomplished using look-up functions available in most computer spreadsheet applications. To help minimize the amount of process data required (especially when using months of lab data), the time stamps for the process data should provide enough granularity so that it can be mapped reasonably well to when the lab sample was obtained. While this mapping is predicated on the posted sample-capture

time in the lab information management system (LIMS), it is recognized that, very often, the actual sample time differs. Nevertheless, it is often the only data point available, so use whatever is available but also be prepared to deal with any time errors that may exist.

So, taking the Lights Removal Tower example to the next step, the only composition analysis data available was lab based with only 1–2 samples per day taken. Consequently, a fairly large time period (36 months) was used in the temperature–composition correlations. As described above, the process temperatures are matched in time with the time that the lab samples were obtained from the process (as provided by the LIMS) and then plotted, refer to Figure 5.7.

Here the correlation between the overheads distillate stream lights composition to each of the tower's temperatures is quite readily seen: the vertical agglomerations (representing temperature indicators [TIs] B, E, G, and H) represent no correlation, whereas TI-D begins to show some correlation with the overhead distillate lights composition. However, the very obvious, strong, and linear correlation with the distillate stream lights composition is to TI-A which, in this case, is the temperature of the overhead vapor stream.

There are two (2) other items to note regarding this example, which relate to this analysis:

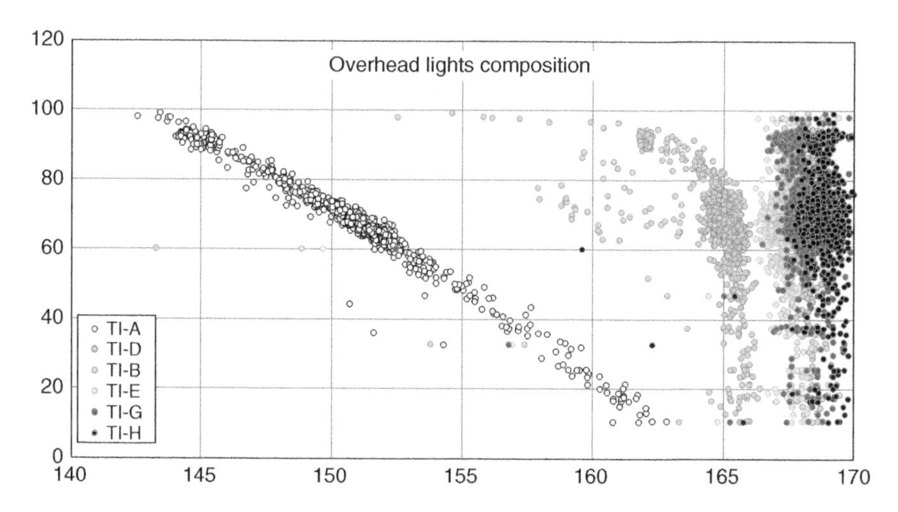

Figure 5.7 Distillation tower overhead composition v. stage temperature.

1. The lights component(s) in the tower feed was a very small fraction, thus the material balance plays a very strong role in the tower's performance. A gain analysis from a plant step test (or validated steady state simulation) would corroborate this assertion.
2. The overheads vapor stream is totally condensed. Thus, the compositions of the overhead vapor and the liquid streams (reflux and distillate) are all identical. This, in turn, allows the overhead vapor temperature (TI-A) to be used as a direct inferential for the distillate stream composition.

The obvious disadvantage of this procedure is that nothing is known about the dynamic response, and hence the dynamic disturbance rejection capability of alternative control schemes is also not known. These need to be evaluated using a dynamic simulator.

5.6 Distillation Control Scheme Design Using Dynamic Models

As detailed above, the steady-state methodology can be used to screen a large number of candidate control schemes quickly and efficiently. However, it is desirable to then evaluate the candidate control schemes using a dynamic simulator to check the dynamic disturbance rejection capabilities of the alternative control schemes. Many case studies that demonstrate the application of this design scheme are available in the literature (e.g. Young and Svrcek 1996) and the procedure is also to be followed in the workshop accompanying this book.

The basic dynamic design procedure consists of the following five steps (which follow upon the steady-state procedure):

1. *Provide information on material hold up:* Material may be held up in the condensers, tray sections, reboilers, and other equipment used in the process. These delays will affect a controller's ability to respond in a reasonable amount of time when attempting to smooth out process disturbances. Typical hold-up times for various equipment items are listed in Table 5.1.

Table 5.1 Typical equipment hold-up times.

Equipment	Hold-up time
Heat exchangers	30 s
Distillation column trays	15–30 s (larger for crude columns)
Distillation column reflux accumulators	5–10 min
Distillation column reboiler/ bottoms accumulators	15–20 min or large enough to hold up liquid from trays (i.e. if dumped)
Surge vessels	10s of minutes to hours

2. *Add controls and instrumentation:* Depict cascade control when necessary, add lags and dead time to process measurement if this is expected.
3. *Implement control structures:* Tune these controllers.
4. *Practice with dynamic simulation:* Test for stability at startup, shutdown, runtime, typical upsets, etc.
5. *Repeat steps 3 and 4 for each control strategy:* Compare ease of startup and shutdown, disturbance rejection, what would happen if there were a change in the rate of production, complexity, and interaction between the controllers (e.g. Mahoney and Fruehauf 1994; Svrcek et al. 2014).

A case study that demonstrates the application of these design approaches is available in the literature (Young and Svrcek 1996; Svrcek et al. 2014) and is also included in one of the workshops, which accompanying this book.

Tutorial and Self-Study Questions

5.1 What are the pre-requisites of using temperature as a proxy for product specification control?
5.2 Describe the characteristics of temperature control loops.
5.3 Describe set up of temperature control loops.
5.4 Describe examples of different temperature loops.
5.5 Describe the purpose of different temperature loops.

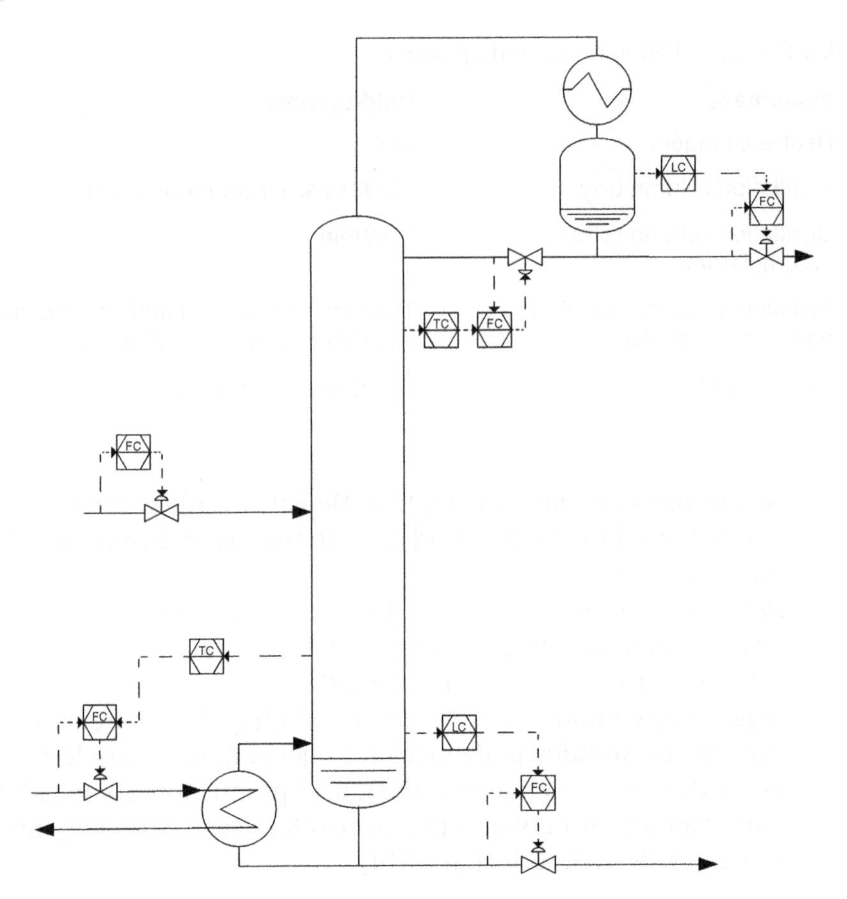

Figure 5.8 Distillation tower with double-ended composition controls.

5.6 Why do reboiler outlet temperature controls sometimes work (that is provide a correlation with composition) and sometimes do not?

5.7 What is going to happen at "high" feed rates to the tower in Figure 5.8? Why will this happen? How do you "fix" it? What else should you check or confirm?

References

Cutler, C.R. and Ramaker, B.L. (1979). Dynamic matrix control – a computer control algorithm. *IEEE Transactions on Automatic Control* 17: 72.

Fruehauf, P.S. and Mahoney, D.P. (1993). Distillation column control and design using steady state models: usefulness and limitations. *ISA Transactions* 32 (2): 157–175. https://doi.org/10.1016/0019-0578(93)90038-X.

Fruehauf, P.S. and Mahoney, D.P. (1994). Improve distillation-column control design. *Chemical Engineering Progress* 90 (3): 75–83.

Mahoney, D.P. and Fruehauf, P.S. (1994). An integrated approach for distillation column control design using steady state and dynamic simulation. *Proceedings of the 73rd GPA Annual Convention*, New Orleans (7–9 March 1994), 72–80.

Prett, D.M. and Gillette, R.D. (1980). Optimization and constrained multivariable control of a catalytic cracking unit. *IEEE Transactions on Automatic Control* 17: 73.

Ryskamp, C.J., "New strategy improves dual composition control (also effective on thermally coupled columns)", *Hydrocarbon Processing*, June 1980, p. 51–59.

Svrcek, W.Y., Mahoney, D.P., and Young, B.R. (2014). *A Real-Time Approach to Process Control.* Wiley.

Tanaka, H., Ohara, T., Ryu, D., and Hopkins, C. (2010). Rapid analysis of gas & liquid phase using NR800 near-infrared analyzer – application to petrochemical process such as ethylene plant and chemical process. *Yokogawa Technical Report English Edition* 53 (2): 55–58.

Michael A. Taube, Isuru A. Udugama (2021). A real-world case study: (re)assessing distillation tower instrumentation & controls for optimal operations. *2021 AIChE Spring Meeting*, Dallas, Texas (April 2021).

Tolliver, T.L. and McCune, L.C. (1978). Distillation column control design based on steady state simulation. *ISA Transactions* 17 (3): 3–10.

Trevedi, Y. (1993). Controlling distillation with the most sensitive tray. *Chemical Engineering* 100 (1) Jan.: 141, 145.

Tyreus, B.D. (1992). Selection of controller structure. In: *Practical Distillation Control* (ed. W.L. Luyben), 178–191. Van Nostrand Reinhold.

Young, B.R. and Svrcek, W.Y. (1996). The application of steady state and dynamic simulation for process control design of a distillation column with a side stripper. *Proceedings of Chemeca '96, 24th Australian and New Zealand Chemical Engineering Conference*, Sydney. Published by the Institution of Engineers, Australia, 01 September 1996, Vol. 2, pp. 145–150.

6 Refinery Versus Chemical Plant Distillation Operations

This chapter focuses on distillation columns in classical refinery settings (as often presented in the control literature) compared to (petro)chemical plant distillation operations. The discussion is based on the authors' experience and knowledge developed over decades of practice. However, for the reader who desires some more background, excellent references are those of Kister (1990) and Kaes (2000) for refinery operations and simulation, respectively.

Due to contrasting economic and quality objectives explained below, the operational doctrine of distillation columns across industrial sectors are somewhat different. Because of these factors, from a process and quality control perspective, the main difference between refineries versus (petro)chemical plants amounts to this: Refinery quality objectives are focused on bulk fluid properties, such as viscosity, boiling points, freeze, or cloud point (which are also subject to seasonal market considerations), whereas (petro)chemical plant's quality objectives are focused on specific molecular species, which may, or may not, include different quality grades (e.g. chemical grade versus polymer grade). The former quality measurements are best determined by lab analysis, while the latter is easily accomplished with online analytic instrumentation (GC, photometric, etc.). The economic consequence of these differences is that refineries can shift molecular species between product streams to maximize volumetric production of the higher margin product(s). That is they can shift molecules from lower margin streams to higher margin product stream(s), while still maintaining the quality objectives. It is this ability to shift molecules between product streams and the always-shifting product margins that have made

A Real-time Approach to Distillation Process Control, First edition. Brent R. Young, Michael A. Taube, and Isuru A. Udugama.

Companion website: www.wiley.com/go/Young/DistillationProcessControl

refinery operations and economics such as *black art* with the resulting "feast and famine" roller-coaster ride! Whereas (petro)chemical plants run at nominally fixed quality and production targets based on product demand.

There is also a historical perspective worth noting: In years past, refinery distillation operations had much "looser" objectives compared to chemical plants. For example, many product specifications were vaguely specified (e.g. having significant slack), and off-spec production was often sent on to product storage where the motto was "We'll 'fix' it in the Blender!" Furthermore, the distillation towers, except for crude towers, were also much smaller in terms of physical dimensions, especially when considering the number of trays/stages. However, as time progressed, petrochemical, as well as specialty chemical, plants were required to meet much more stringent (and consistent) product specifications and, particularly, lower impurity specifications (e.g. parts-per-million range).

The inevitable outcome of various quality philosophies (such as Statistical Quality Control [SQC] and tighter capital budgets) forced more stringent product specifications upon refineries, which supply feedstocks to the downstream sectors. The net result of these market forces is that refineries are now required to operate more like chemical plants. Unfortunately, refinery unit designers and operations engineers have yet to grasp what this means and its effect on the design and operations, as well as requisite control structures, for refining distillation columns.

While refinery product specifications are still "low purity," by chemical standards – in the <10% molar volume range – in comparison to their historical limits, refineries now find themselves operating as "high purity" distillation facilities, as follows:

1. Refinery distillation operations are more energy-intensive: Reflux/Distillate (R/D) or Reflux/Feed (R/F) Ratios are larger (>2) compared to historical values (<1–1.5).
2. The towers themselves are physically larger with more trays/stages, thus dynamic characteristics are also larger (longer) and behave more like chemical units.
3. The use of online analyzers, such as gas chromatographs (GC) and/or photometric analyzers, for closed-loop control, in order to facilitate faster feedback correction to temperature and inferential-based controls, is increasing.

While from a mechanical design, VLE/theoretical stages and operations perspective, refineries are looking more like high-purity distillation facilities, refinery control designs still resemble those of "low purity" or "classic refinery" style, as shown in Figure 6.1a. As illustrated, the distillation column has only a single stream specification: the lights fraction leaving with the bottoms stream. As indicated by the tray temperature used to set reboiler duty; the overhead stream composition is kept "open" due to the column's overall mass and energy balance not being fixed. As such, any disturbance entering the column is rejected by the bottoms stream specification to the distillate stream, which absorbs these disturbances.

Over time, however, the economic impetus to recover the "lost" bottoms product increased, which meant that a second controller to minimize or recover the heavy component in the overheads stream is added. Without considering the real-time operational impacts, the engineer responsible for the unit utilizes the only open degree of freedom (DoF) – the reflux flow – as the MV for affecting the tray temperature that correlates with the heavies composition in the distillate stream. The resulting control structure is shown in Figure 6.1b.

From Chapter 2, one should recognize this control structure as the LV type of control structure. That is the reflux flow (L) is used as the manipulated variable to control the distillate composition, while the reboiler duty (influencing the vapor boil-up rate, V) is used to control the bottoms composition.

To understand the inherent issues such a set of control structures may cause, let us conduct a thought experiment: Let us assume that the concentration of heavy impurities in the distillate stream is high. The top composition controller will detect this in the form of a temperature increase at the top of the column which is above the desired setpoint. To remedy the situation, the top composition controller (L controller) will increase the reflux flow rate. With this higher reflux flow, the temperature at the top of the column would go back down towards the set point. However, as a direct result of this increased reflux, the temperature on each column tray, starting from the top, will reduce. One can think of this as a "cold front" propagating toward the bottom of the column. Once this "cold front" arrives at the bottom of the column, or more specifically, the tray at which the bottoms temperature is monitored, the bottom composition controller (V controller) detects that the bottoms temperature is dropping lower than the desired setpoint. To get

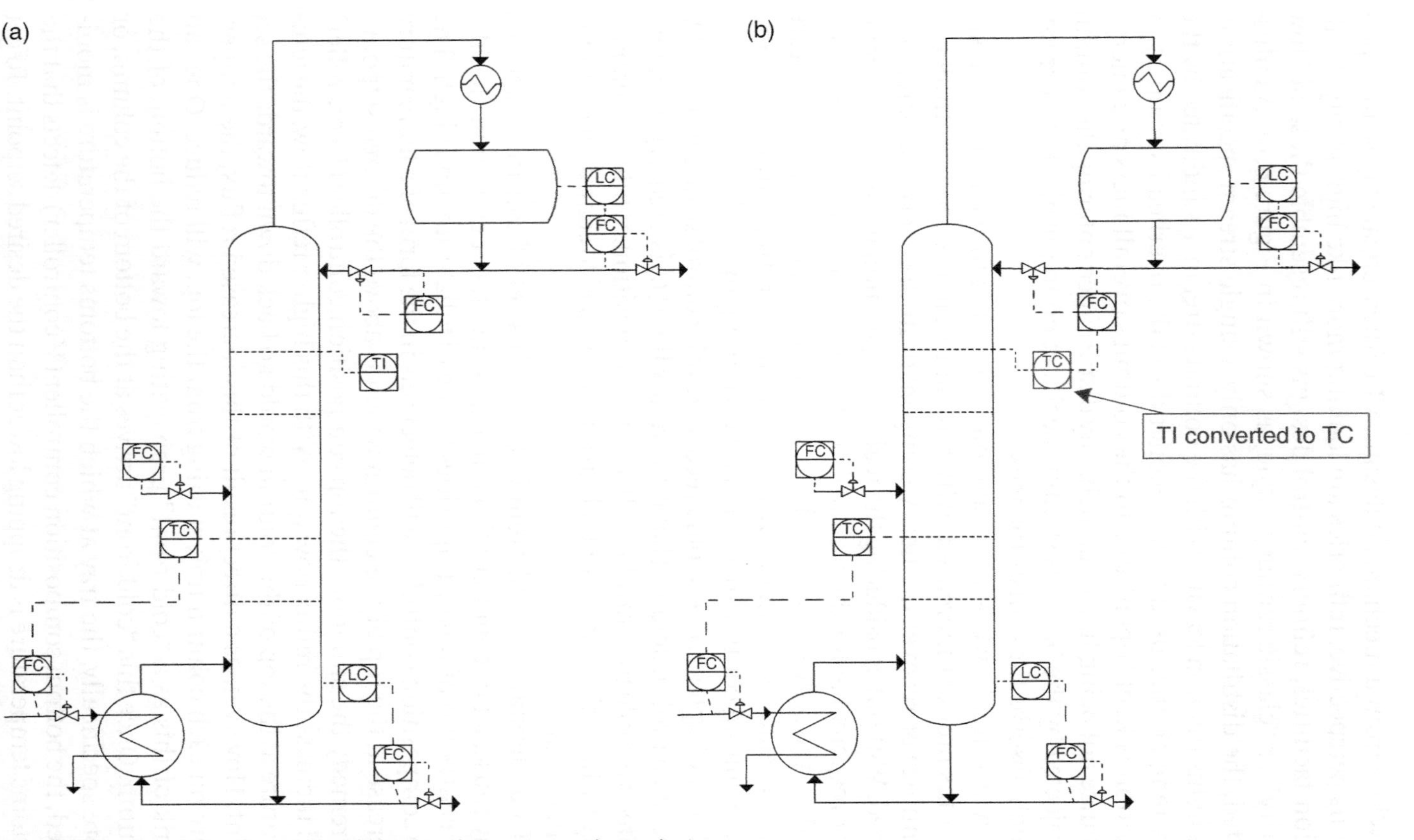

Figure 6.1 Typical refinery distillation column controls evolution.

the temperature back up to the desired setpoint, the V controller will increase the boil up of the column, which would create a "hot front" that will propagate up the column. And thus, the cycle repeats itself.

From a control point of view, this type of cycling behavior is undesirable and leads to one of the following outcomes.

- Flooding: In a situation where the column is operated close to its rated mechanical capacity (which many industrial columns are due to countless capacity improvements projects that are carried out over the lifespan of a facility). These "hot and cold fronts," where there is an excess of liquid and vapor traveling through the column, can easily (and usually does) result in flooding. Note: flooding can happen because of both increased liquid and/or vapor loading. Once flooding has set in, the inherent separation loss will ensure that the two controllers keep on "pushing" the column further into flooding. Only operator-intervention can properly address the situation, where the remedy is to reduce the feed flow to the column or to "break" the cycle by "dumping" the column (e.g. sending most of the feed out the bottoms until the tower settles out). As a result, operators are reluctant to operate the column close to the "maximum capacity" of the column.
- Cycling: Any change in external conditions or feed compositions will send this type of control structure into a "cycle" mode, even when operating well below mechanical limits. The amplitude and period of these cycles depend upon the size of the tower and the relative "distance" (thermodynamically) between the component boiling points, as well as the controller tuning (e.g. aggressive v. loose).

A principal reason for these type of "upgrades" being commissioned is the general lack of understanding by process engineers on how to affect distillation columns: Per Chapter 2, there are only two (2) "handles": mass balance and energy balance. The two (2) controllers shown in Figure 6.1b – one for the bottoms composition and the other for the distillate composition – are using the SAME HANDLE, energy balance – to achieve their objectives. Thus, as, say, feed is increased, both controllers will increase their MVs – reboiler duty and reflux flow – and, eventually, flood the tower. This behavior is much like squeezing an oblong balloon from both ends!

Another one of the key reasons for these "oversights" is the process engineer's tendency to think in terms of "steady state," and ignore the dynamic behavior. Consequently, they fail to grasp how the controls will interact in real time.

6.1 New Generation of Refinery Controls

As stated earlier in this chapter, most refineries today have similar operational requirements to chemical plants. Which is why many refineries have attempted to achieve their new operational goals with incompatible control structures and stop-gap minor upgrades, but with limited success. To this end, today's refinery operations need to adopt more of the chemical plant-based control structures that were discussed in the Chapter 5.

There are, however, refinery-specific considerations that must be taken into account when developing controls for modern refineries to wit: Unlike typical distillation operations, as found in petrochemical and chemical plants, crude oil atmospheric and vacuum towers lack a true reboiler for two (2) fundamental reasons:

1. The very high temperatures at which the bottom of an atmospheric crude tower operates means a reboiler can be technically and economically difficult to justify, due to fouling, as well as required heat input.
2. Introducing (additional) heat at the reboiler may result in "thermal cracking" of heavier compounds in the crude oil mixture, which is also undesirable.

Based on this, a feed pre-heater is used to provide the necessary energy-input to operate the column. Also, where additional heat or vapor flow is needed within the tower, direct steam injection is the preferred mechanism, especially near the tower bottoms where lack of fluidization will result in solids deposition (i.e. coking).

To help distinguish oil refinery crude units from typical distillation operations, the authors recommend this concept: A crude oil tower is, in effect, a flash drum with trays and/or packing. Thus, the feed heaters are providing the energy to affect the flashing of the crude oil as it enters the tower. As there is excess energy put into the feedstock, to affect the separation of the lighter components, side draws from the tower are utilized as heat sources by other areas or units within the refinery.

The cooled liquid streams are then returned to the crude towers in lieu of "reflux." This topic covered further below.

In terms of control, referring back to the discussion in Chapter 4 regarding reboiler controls, the objective with Crude Unit Feed Heaters is simply to increase the enthalpy of the incoming feed stream. Once the feed flashes inside the tower, other MVs are used to affect the quality of the product streams: primarily side-draw flows and, to a lesser extent, the pump-around flows. As fuel gas usage in the feed heaters is a major economic handle for Crude Oil Units, the amount of heat input is the main handle for energy optimization.

However, unlike a reboiler, the changing of the heater duty directly affects the vapor/liquid split between the rectifying and stripping sections of an atmospheric crude tower, which must be well managed to ensure reliable operations. Also, depending on the feed composition (i.e. "light" crude v. "heavy"), increasing the preheater duty could result in an overloading of the rectifying section, which might result in flooding, while leading to the "drying out" and/or "coking" of the stripping section as the liquid that would otherwise flow down the tower to the bottom is now vaporized. Conversely, reducing the preheat duty (feed enthalpy) allows lighter components (which are typically valued higher than heavier components) to remain liquid and go out with the heavy residual oil bottoms stream (e.g. an economic loss). Thus, as with many other energy-intensive unit operations, there is a balance that must be maintained between energy input and quality and operational objectives.

6.1.1 Atmospheric and Vacuum Refining Columns

Compared to traditional chemical distillation columns, there are some nuances related to the design and operations of both atmospheric and vacuum crude oil refining towers, which are not typically found in (petro)chemical plants. These nuances mainly come in the form of pump arounds and side strippers.

6.1.1.1 Pump Arounds

Pump arounds are used for two (2) primary purposes:

1. Removal of excess heat (e.g. heat integration with other refinery units); and
2. Manage vapor/liquid loading within portions of a Crude Tower to affect molecule selection of the product side draws.

Operationally, the pump around draw is taken from tray/stage, pumped over to the "heat sink" (i.e. one or more heat exchangers in other units that use the excess heat), and then returned back to the course crude tower a few trays/stages above the place from which it was drawn. The subcooled liquid then interacts with the vapor coming up from the tray/stage below, as well as mixing with the liquid coming down from above in the tower, as illustrated in Figure 6.2. Ultimately, the pump around provides a degree of freedom for affecting fractionation within that portion of the crude tower from where it was drawn and to where it was returned as "reflux." When coordinated with other MVs, the distribution of molecules between the side draw streams is optimized to generate maximum profit for the refinery. Thus, there will be multiple factors (or controls) which specify the pump around flowrate, recalling that the purpose of all Crude Tower controls is to influence the selection of molecules leaving via the product draws to maximize (or minimize) the volume produces while meeting all of the stream quality specifications. But we were getting ahead of ourselves now!

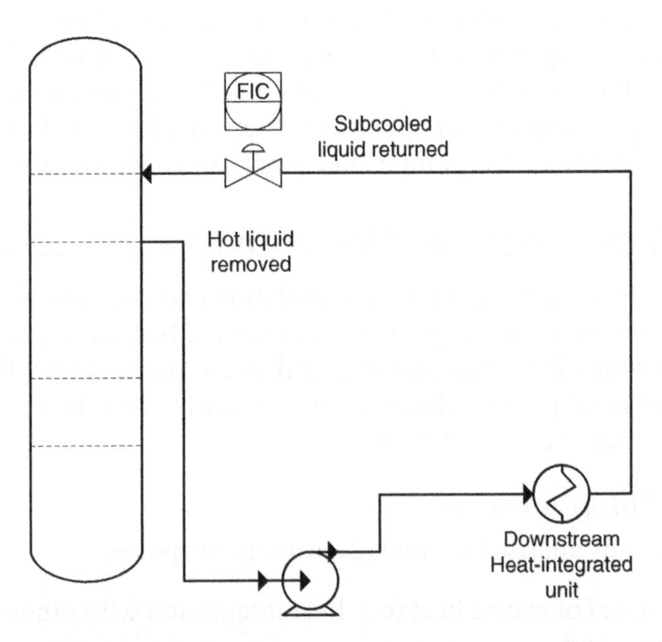

Figure 6.2 Pump around controls.

6.1.1.2 Side Strippers

In decades past, side strippers were used to remove lighter components from a side draw product. This was achieved by pumping a side stream from the main tower into a side stripper vessel where, usually, direct steam injection was used to evaporate the lighter components while leaving the heavier component(s) in the liquid phase, as shown in Figure 6.3. The flow into the side stripper is determined by the temperature near the side draw of the main column, where a higher concentration of the product to be removed will result in a higher flow rate to the side stripper. The steam injection was used to maintain the temperature in the side stripper such that only a concentrated product remains in the bottom's flow. Over time, however, many of these vessels were abandoned or reduced to being a "wide spot" in the line with no steam injection as quality control improvements and minimizing utility usage efforts were pursued.

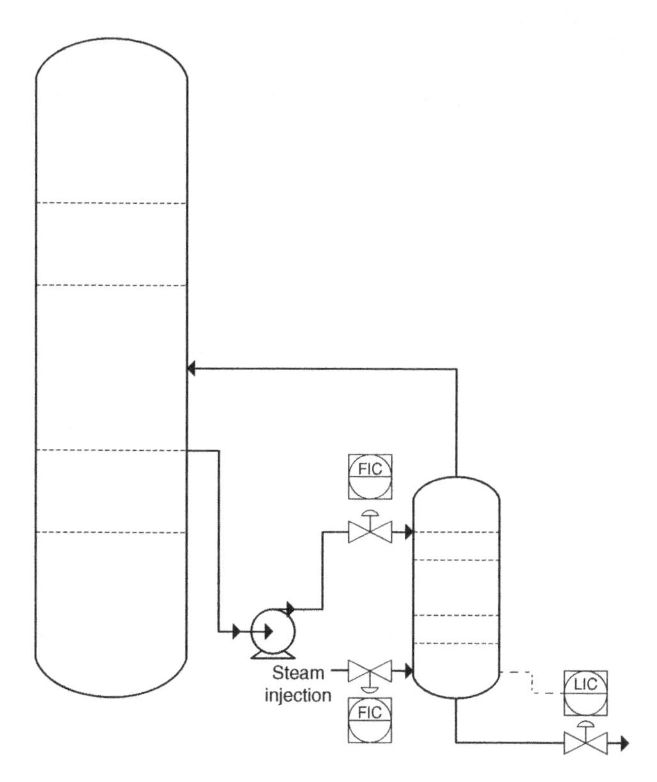

Figure 6.3 Side stripper controls.

6.2 Improving Thermodynamic Efficiency Through Control

In a refinery, as with all other process facilities, managing the utilities usage is an important task. Significant heating and cooling processes are taking place where operating temperature can reach greater than 350 °C in a fired heater to less than −150 °C in cryogenic separation processes. To this end, the efficient control of heat and cooling resources is required for economical operation. In terms of distillation operations, the overall control philosophy might require reconsidering the typical concepts.

For example, at the extremities of the operating temperature range, it might be more economical (hence optimal) to conserve/manage energy usage than ensuring all feedstock is processed. This is different from the conventional operating wisdom where the energy source (fired heater/reboiler/cryogenic cooling) is adjusted to ensure all feedstock available is processed. However, with the high economic cost of providing heating/cooling duty, the feedstock may be varied such that only the available heating/cooling duty is used.

6.3 Blending and Its Implications on Control

Blending is mixing multiple product streams to create a petrochemical product that meets a given market specification (for example 95 octane petrol). The practice of blending also means that a slight variation in stream specification will not affect the final product specifications as long as the other streams in the blending pool can counter these variations (i.e. "fix it in the Blender!"). Hence, in operations where heating/cooling costs are high (high or low-temperature processes), the product stream specification can drift with a pre-defined range rather than changing a reboiler (condenser) duty or feed flow rate. Thus, any variations in this stream can be compensated for in blending by having a tighter control on the product stream where the operations have a lower cost of heating/cooling.

Tutorial and Self-Study Questions

6.1 If the vapor traffic in a refining column section is increased, how would a pump around detect and control this? What are the consequences of such action product purity?
6.2 What are the differences between refining and (petro)chemicals regarding quality measurements?
6.3 Alternatively, what are the differences between refining and (petro) chemicals regarding operational (control) objectives?
6.4 How have refinery operations been affected by end-user quality programs over the last 20+ years? What does this mean for refinery designers and operations support (process and process control) engineers?

References

Kaes, G.L. (2000). *Refinery Process Modeling*. Elliott & Fitzpatrick.
Kister, H.Z. (1990). *Distillation Operation*. McGraw Hill Professional.

7 Distillation Controller Tuning

The behavior of a process control loop can be completely changed by its tuning and is often the difference between stable and unstable operations. As such, a key aspect of implementing any control structure is to ensure that a given control loop is tuned to achieve the unit's (or equipment's) operational objective. Many academics and industrial practitioners have worked in this topic and have introduced and continue to introduce tuning rules that can be applied to a process by gathering information-based open-loop or closed-loop testing (e.g. Yuwana and Seborg 1982; Åström and Hagglund 1984; Tyreus and Luyben 1992). Zeigler-Nichols (1942) is an early closed-loop method that is referred to extensively (but should NEVER be used). A controller's process variable (PV) is forced into a stable and continuous oscillation loop by adjusting the tuning to obtain information about the process' characteristics. The information gathered is then employed to tune the control loop. However, Zeigler Nichols tuning was developed for an entirely different purpose that controlling distillation columns is completely UNSUITABLE for distillation operations and any other unit operation in the hydrocarbon processing industry. More nuanced approaches, such as lambda tuning (McMillan and Vegas 2019), as well as improved variations of IMC (internal model control) tuning (Rivera et al. 1986; Skogestad 1983; Fruehauf et al. 1993), have also been developed more recently. This chapter aims to illustrate how process identification and how controller tuning should be performed for distillation unit operations. In turn, it will be shown how the controller behavior (resulting from tuning) is used to achieve the overall operational objectives for a distillation column.

A Real-time Approach to Distillation Process Control, First edition. Brent R. Young, Michael A. Taube, and Isuru A. Udugama.

Companion website: www.wiley.com/go/Young/DistillationProcessControl

7.1 Model Identification: Step Testing

The first step in every tuning method is to identify both the system's underlying process dynamics and steady-state behavior. For many tuning methods, this is determined from an open-loop step test. The data is then used to fit an assumed process model. While the data can be correlated to complex dynamic process models, a simple First Order + Dead Time (FO + DT) is the most common and provides an acceptable fit for many, if not most, processes. A PID controller has three tuning variables – K_c (controller gain) and the integral and derivative terms (usually expressed as time variables). Hence, if a process is described by the three variables of K_p (Process Gain), Lag Time and Dead Time, then these variables can be easily mapped to the three parameters in the PID controller. While more complex process models will have additional variables (for example a second lag or a lead), the process model must still be mapped to the three PID tuning variables somehow.

Figure 7.1 is a simple illustration of an open-loop step test. Using this example, the K_p (process gain), lag time, and dead time are determined as follows:

- The dead time (θ) is the time before a movement is detected in the process variable.
- The lag time (τ) is the time taken for the process variable to reach 63.2% of its final steady-state value.

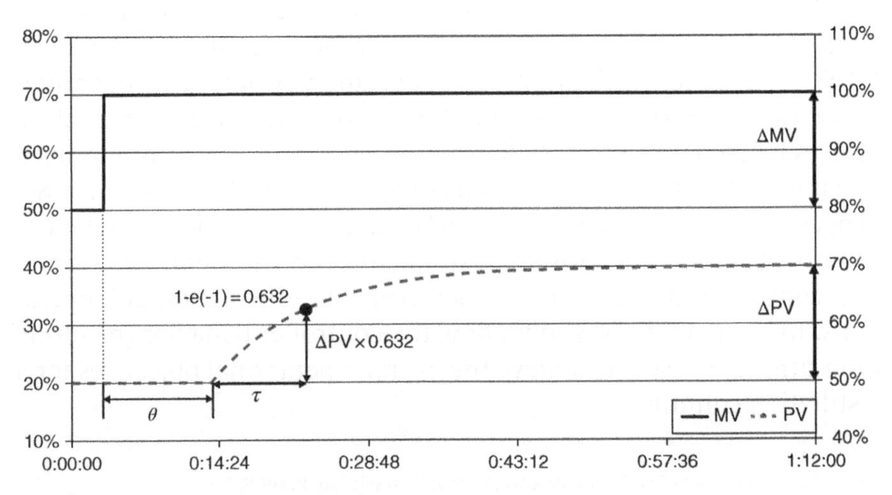

Figure 7.1 First-order plus dead-time process response.

- The process gain (K_p) is determined by dividing the change in CV (or process variable, PV) by the change in MV (or controller output, OP), as shown in Eq. (7.1).

$$\frac{\Delta \text{PV}}{\Delta \text{OP}} = \frac{\Delta \text{CV}}{\Delta \text{MV}} = K_P \tag{7.1}$$

The Engineering Unit values should be converted to Range-Based or Percent-of-Scale values to normalize the values for different MV–CV pairings. This is done by dividing each formula term by the range of the associated instrument, as shown in Eq. (7.2), where the *Max* and *Min* designations refer to the respective instrument ranges.

$$\frac{\Delta \text{CV}/\left(\text{CV}_{\text{max}} - \text{CV}_{\text{min}}\right)}{\Delta \text{MV}/\left(\text{MV}_{\text{max}} - \text{MV}_{\text{min}}\right)} = K_{P_{Range-Based}} \tag{7.2}$$

"Instrument Range", in this instance, refers to the calibrated range of the transmitter (relevant to the PV/CV and the OP/MV; if the MV is a secondary controller in a cascade scheme). In the case of a control valve, the flow capacity of the valve is from 0% to 100% opening. Note that ***Max*** and ***Min*** will often be values other than 100 or 0, respectively, particularly for temperature controls. So always verify the calibrated range!

This information can now be used with many tuning methods to get a set of K_c (controller gain), integral, and derivative values for a PID controller. Note: controllers in commercial environments cannot be set to have a negative gain. Instead, the controller can be set up to be a reverse or a direct-acting controller. In practice, noise/uncertainty are inherent in process measurements, as illustrated in Figure 7.2. Attempts to fit process models to something more complex than a simple FO + DT are likely to suffer from overfitting.

7.2 Typical Process Responses

The open-loop process responses observed in chemical processes fall into two categories. These are open-loop stable processes and open-loop integrators. *Open-loop stable (OLS) processes* are processes that achieve a new steady-state value after making a change in the MV such

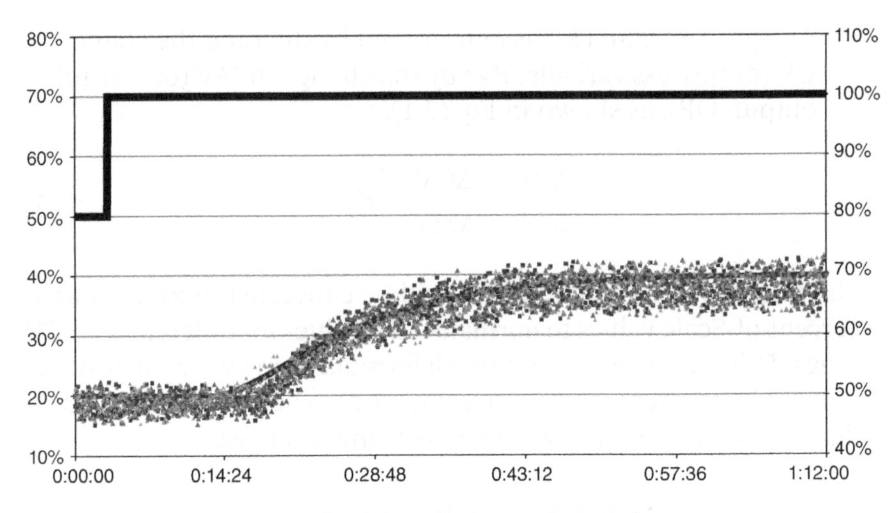

Figure 7.2 First order plus dead time fit to actual plant data.

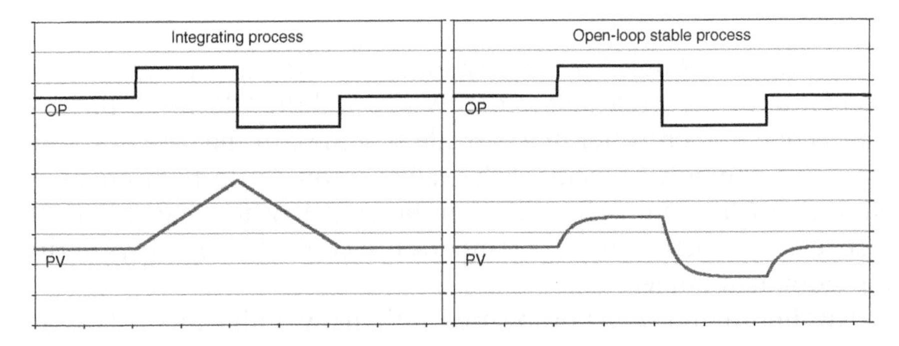

Figure 7.3 Integrating versus open-loop stable process step responses.

as flows, temperatures, and stream compositions. This is compared to an *integrating processes*, such as a tank or drum level, that never reaches a steady state, as illustrated in the two graphs in Figure 7.3.

The derivation for calculating the *process gain* (K_p) for a tank or drum level was given in Chapter 4 so will not be repeated here. Nevertheless, it must be noted that one is likely to find other types of process measurements, besides levels and pressures, that act like an *integrating process* (e.g. some distillation tower temperatures), as well as some levels and pressures that behave like *open-loop stable processes*! So be vigilant to the mechanical design and always confirm with an open-loop step test how the process actually behaves. Once this is confirmed, then appropriate tuning rules may be applied to give the closed-loop responses shown in

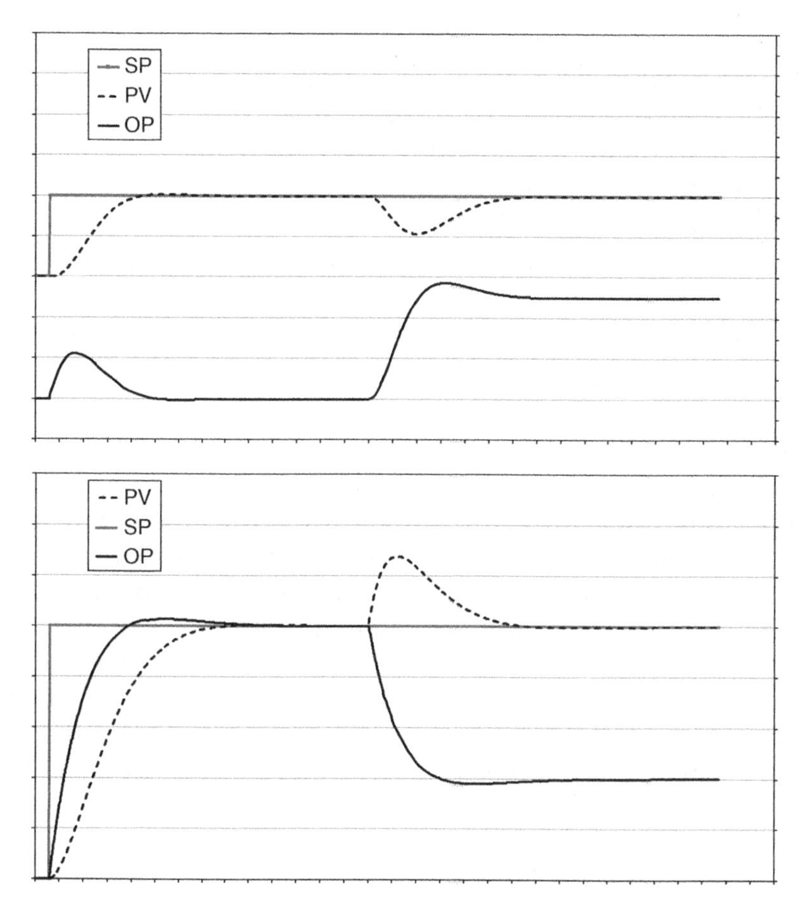

Figure 7.4 PID controller responses for integrating versus process open-loop-stable.

Figure 7.4 for both types of process behaviors. Note that these controller responses are proposed as being optimal for all process industries unit operations' controllers.

7.3 Engineering Units Versus Percent-of-Scale

Several times it has been mentioned that one should obtain Percent-of-Scale values for the *process gain* (K_p) rather than simply using Engineering Units. There are three (3) reasons for obtaining a *Percent-of-Scale* value:

1. To normalize all the possible MV–CV pairings, so they can be evaluated against each other on a common basis, particularly if the engineering units for the MVs differ.
2. It clearly indicates how much MV movement is required to respond to a disturbance on a percentage basis; this aspect will be explored in more detail, below.
3. Many control systems use *Percent-of-Scale* values for the *controller gain*, (K_c), so it helps to have the *process gain* already in *Percent-of-Scale* units so as to generate the required *controller gain* when it is time to commission the loop.

In order to explore how much MV movement is required to address or mitigate a process disturbance – whether measured or unmeasured – the *Percent-of-Scale* form of the *Process Gain* calculation provides a very clear indicator: in short, the *inverse* of the *Process Gain* indicates how much movement of the MV, on a percentage basis, is required to respond to a 1% change in the CV *at Steady State*; the controller's dynamic response may be larger or smaller, but once the dynamics have passed, all that remains is the new steady-state value. Understanding the importance of this facet first requires that one examine the relationship between the *Controller Gain* (K_c) and the *Process Gain* (K_p), as given by Eq. (7.3).

$$K_C = \frac{K}{K_P} \tag{7.3}$$

The *K* term in Eq. (7.4), referred to as the *loop gain*, is obtained from tuning rules based on the type of process (OLS versus integrating), process dynamic characteristics, the structure of the PID algorithm, and the control objective function (i.e. the desired response to SP changes and unmeasured load disturbances). For the purposes of this evaluation, the *loop gain* is ignored, since the key factor here is determining the final steady-state change that this MV–CV pairing requires to respond to a process disturbance. This response is illustrated by plotting the *controller gain* (K_c) as a function of the *process gain* (K_p), as shown in Figure 7.5. This very nonlinear relationship clearly illustrates the importance of proper MV-CV pairing: As the *process gain* becomes smaller, the *controller gain* (or the steady-state response) will

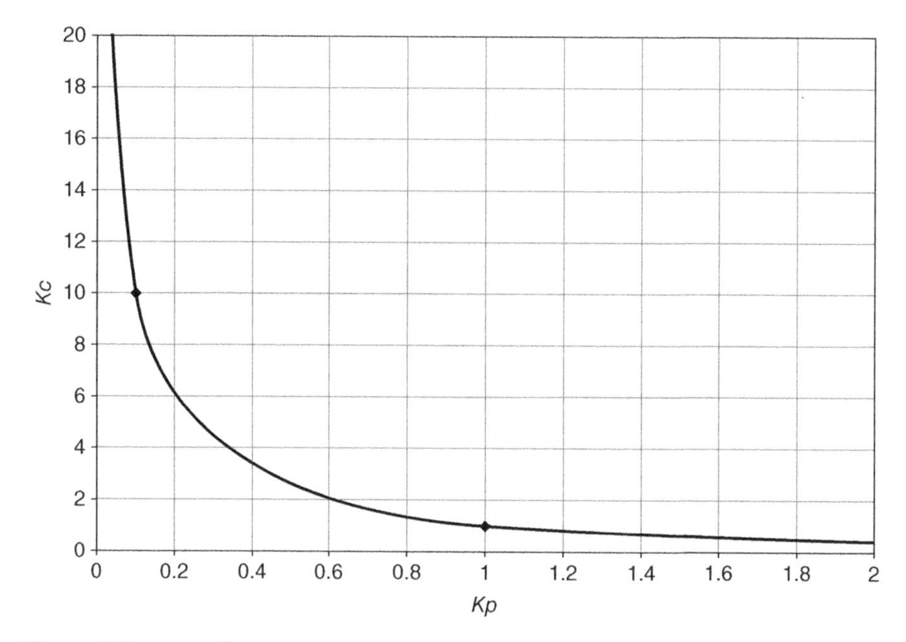

Figure 7.5 Controller gain versus process gain.

grow exponentially larger! Note that this is true for BOTH *OLS* and *integrating processes*.

The consequence of getting the MV–CV pairing wrong or poorly matched is illustrated in the graphs in Figure 7.6. The plots shows a level controller response for both a set point and unmeasured load change for a typical drum level measurement as one might find in a running plant, including "noise." Figure 7.6a illustrates a good MV–CV Pairing, while Figure 7.6b shows a poor one. The only difference between the two simulated controller responses is the *process gain* (K_p): that is one is using a larger flow capacity MV than the other. There are two (2) observations one should make regarding the second plot:

1. The controller OP is "maxing out" its value to 100% (the scale for the black trace is actually 0–200% so as to keep the traces from overlapping). So, the PV will continue to decline into the future.
2. The PV trace in Figure 7.6b is noticeably less noising compared to that in Figure 7.6a. The reason for this seeming discrepancy is left as an exercise for the reader!

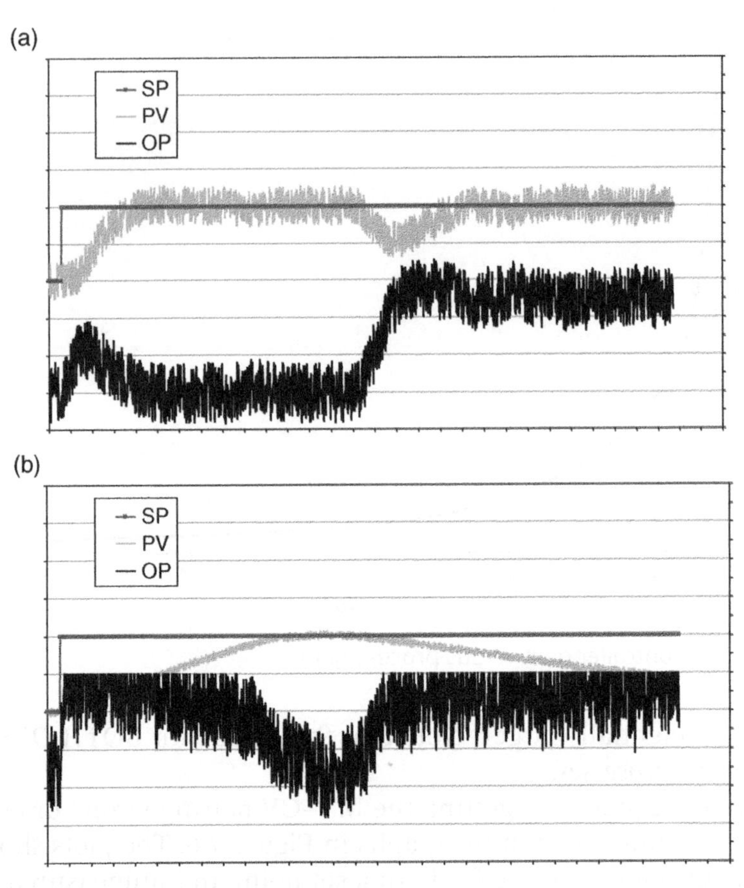

Figure 7.6 Good v. poor MV–CV pairing.

7.4 Basics in PID Tuning

While PID tuning, per se, is beyond the scope of this book, a few tips or guidelines are warranted, as follows.

a. Not all PID implementations are the same! The original pneumatic PID controllers were based on a mechanical design that had the derivative function act as a filter to the other two (2) functions, refer to Figure 7.7a. Hence the designation as the *Interactive PID Form*. The early electronic controllers, as well as distributed control systems (DCS), replicated this form, ostensibly so that the tuning constants could be copied directly from

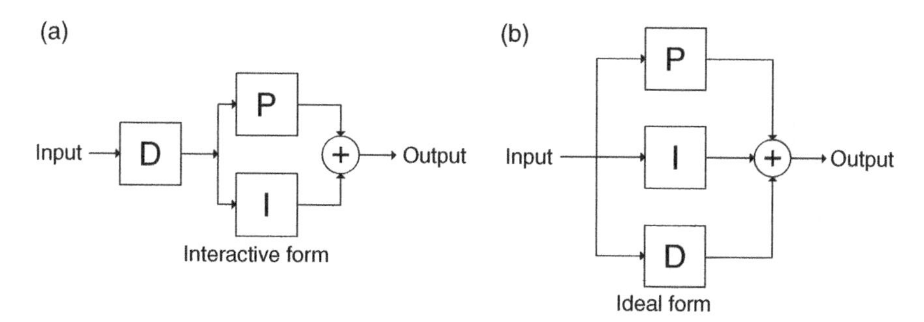

Figure 7.7 Interactive versus ideal PID form block diagrams. (a) Interactive form (b) ideal form.

the pneumatic controller to the new electronic/DCS controller without any conversion. Only after the widespread use of computer-based simulations did the form of the PID algorithm get revised to what is referred to as the Ideal PID Form, refer to Figure 7.7b. As may be elucidated from the block diagrams in Figure 7.7, the tuning for these two (2) different forms of the PID are very different whenever derivative action is required. Thus, when attempting to tune a PID controller, it is vital to know WHICH *PID form* is available on the control system, particularly when derivative action is needed.

b. Regardless of which PID form is used, the "objective function" used by the tuning rules should result in the controller response illustrated in Figure 7.4. These responses are submitted as satisfying the *Purpose of Process Control* as given in Section 1.1. That is, keeping the process at the specified target with the least amount of movement or energy input.

7.5 Tuning in Distillation Control

Generally, it is desired to achieve on target or "tight" control of each controller PV/CV. The main consideration goes back to the opening preface of this book: Maintain the target value using the least total and incremental energy input to the process.

If the product specification(s) is more flexible and the heating/cooling duty for the column is costly, then a more sluggish controller response may be implemented. The main difference between these two approaches is how the controller would "close" the error between the

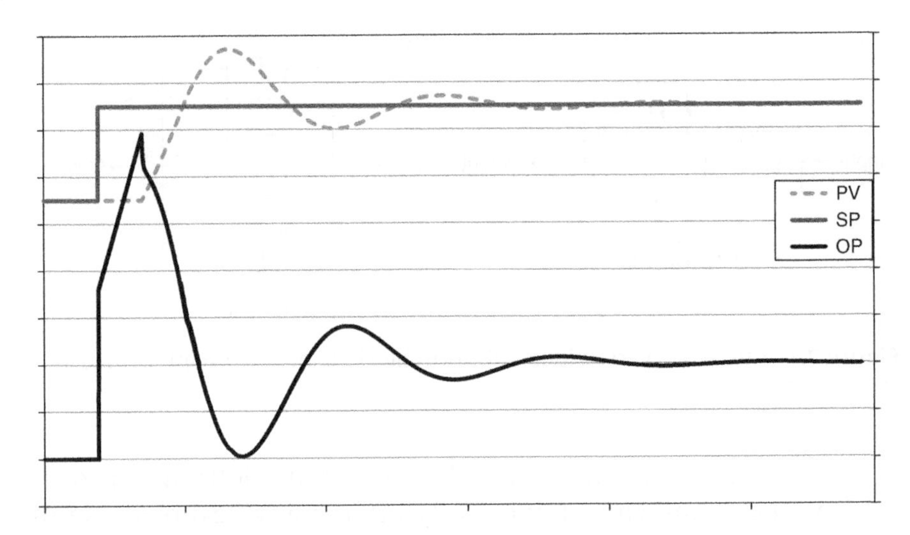

Figure 7.8 QDR controller response.

set point and process variable in response to a disturbance or target (SP) change. The sluggish controller, for example, will take more time to close the error by changing the manipulated variable slowly. In contrast, aggressive tuning will move the manipulated variable rapidly and, perhaps, with "large" changes to get the process back at the set point as quickly as possible. Classically, this trade-off is "optimized" mathematically by forcing the process to respond with a Quarter Decay Ratio (QDR), as shown in Figure 7.8. However, this type of response amounts to a "mathematical fantasy" that does not account for the physical effects to the process as a result of such behavior. As shown in Figure 7.8, the energy input required to achieve such behavior exceeds three times the steady-state energy input. A more overdamped controller response (as illustrated in Figure 7.4) is much preferred as there is no oscillatory movement in the process variable, while the manipulated variable movement is also minimized.

From a technical point of view, there are also future reasons for preferring the "ideal" controller tuning. The first reason for implementing such a response is due to inherent issues in model identification. In any controller tuning method, a process model is identified initially upon which the PID tuning variables are determined. However, in practice, the identification of such a model is affected by the amount of process noise that is present, as well as the ability (or the Operator's agree-ability) to move the controller output for an open-loop step test. At the same time, slight changes to a feed or product profile can easily change the process model identified (i.e. the relationship between the MV and CV changes). In both these scenarios, a controller tuned to follow the QDR based on the model identified can become unstable. In contrast, the "ideal" tuning is better able to handle "model mismatch."

7.6 The Role of Tuning in a "Value Engineering" Era

Traditionally, distillation columns (along with other unit operations) were built with "excess" capacity. This was the result of adding various margins incrementally applied during the process design phase. However, with concepts such as "value engineering" being introduced, such "hidden capacity" is no longer built into a new plant. As a result, a "capacity margin" of, perhaps, only ~5% is likely to be available. At the same time, control and, especially, instrumentation are also reduced to what is deemed "necessary" for the process. One may wonder how these management decisions affect controller tuning. Let us use a simple distillate composition control loop, where a temperature of a stage near the top of a column is used as the quality variable, to answer this question.

Suppose a disturbance hits an older generation column with this type of a control loop. In that case, the controller can be tuned to move the manipulated variable relatively aggressively. This means the temperature (and the composition, by proxy) is maintained at the desired set point. Since there is excess capacity in the column's design, the column can handle the momentary change of vapor/liquid traffic generated by moving the MV. In contrast, in a newer generation column designed with "value engineering," the same control loop and tuning will likely yield a different result, assuming it even has the same room to move. In this instance, as the control loop responds to the disturbance by adjusting the tower's vapor/liquid traffic, it might put the column into flooding, in the worst case, or, lacking the flexibility in MV movement, may result in an extended time of off-spec production.

In addition, the "economization" of instrumentation and control due to value engineering concepts often results in.

- Valves with highly nonlinear %open vs flow characteristics.
- Lack of secondary temperature measurements to confirm sensor accuracy.
- Measurements with a reduced detection range or accuracy (e.g. level measurements that only operate after 30% of a reboiler sump is full, cheap temperature sensors with large sensor detection accuracy).

Ideally, these issues should be dealt with by finding a middle ground between the current "value engineering" practice and the previous 30% + safety margin practice. Thus, the following actions are recommended to manage some of the issues brought on by this practice.

- Dynamic simulations of anticipated disturbances should be run to identify "choke points" in the process design that prevents acceptable response or mitigation. The "choke points" might include maximum flow capacity in process or utilities streams or dynamic vapor/liquid capacities of distillation trays or packing.
- Ensuring that control valve and measurement device designs account for turned-down conditions, as well as having sufficient upward range above the design point.
- Engaging process control expertise much earlier in the process and mechanical design phase than is normally done. This helps ensure that potential design bottlenecks and/or control limitations are identified and addressed before final equipment mechanical designs are issued.

One final point to consider is this: It is well known, but, perhaps, not so well published, that for a greenfield or grassroots plant project, the entirety of the control and instrumentation effort (including, instrumentation, control systems hardware, bulk materials, engineering, and installation) comprises less than 2% of the entire project budget. Thus, any increases to this portion of the project budget would not even show up in the "noise" of the final project cost!

Tutorial and Self-Study Questions

7.1 Two controllers with similar tuning are exposed to a similar set point change: one controller responds by aggressively manipulating the MV to correct the error, while the other responds over time to correct the error. What key difference exists between these two controllers?

7.2 From a very noisy set of step test, you have identified some key system parameters. What precautions should you take in determining PID tuning parameters to ensure smooth operations?

7.3 What are tuning criteria? How are they used? Give some examples, both qualitative and quantitative?

7.4 How would you tune a controller that is current in operations?

7.5 When/where was Ziegler-Nichols published and what was it used for?

7.6 What distinguishes a first-order process from a second (and higher)-order process?

7.7 Why do we always try to fit a process response to a first order + dead time (FO + DT) model? What about for higher order processes?

7.8 What is the purpose of each of the modes or actions – proportional, integral, and derivative – of the PID controller?

7.9 What is the order of importance (1 = Most Important) for accuracy in process model development and identification (i.e. which of the three parameters – Process Gain, Lag Time or Dead Time – is most, second and least important)?

7.10 What is the difference between proportional-on-error (PoE) versus proportional-on-PV (PoPV) control action? What are the consequences of using one versus the other and how does it affect PID controller tuning?

7.11 What is the difference between the positional versus velocity calculation method for the PID controller? When and how does it affect tuning? Why?

7.12 Where (on what kinds of control structures) is proportional-on-PV required and why?

7.13 A PID (SISO) loop is oscillating; what behavior distinguishes proportional-dominant versus integral-dominant tuning?

7.14 What PID controller OP v. PV pattern clearly indicates a sticking valve?

References

Åström, K.J. and Hagglund, T. (1984). Automatic tuning of simple regulators with specifications on phase and amplified margins. *Automatica* **20**: 645.

Fruehauf, P.S., Chien, I.-L., and Lauritsen, M.D. (1994). Simplified IMC-PID tuning rules. *ISA Transactions 33*: 43–59.

McMillan, G.K. and Vegas, P.H. (2019). *Process Industrial Instruments and Control Handbook*, 6e. McGraw-Hill.

Rivera, D.E., Morari, M., and Skogestad, S. (1986). Internal model control, 4. PID controller design. *Industrial & Engineering Chemistry Process Design and Development* 25: 252.

Skogestad, S. (1983). The best tuning rules in the world. *Journal of Process Control* 13 (4): 291–309.

Tyreus, B.D. and Luyben, W.L. (1992). Tuning of PID controllers for integrator/deadtime processes. *Industrial and Engineering Chemistry Research* 31: 2625.

Yuwana, M. and Seborg, D.E. (1982). A new method for on-line controller tuning. *AIChE Journal* 28: 434.

Zeigler, J.G. and Nichols, N.B. (1942). Optimum settings for automatic controllers. *Transactions of the ASME* 64: 759.

8 Fine and Specialty Chemicals Distillation Control

So far in this book, the focus has been on controlling "normal" distillation processes. More specifically, distillation processes that

1. Split the feed composition into distillate and bottoms streams (e.g. "binary" distillation)
2. Involve product purities that are measured in mass % or mole % (e.g. 95 mol% purity)

In the refining operations chapter, the focus was shifted briefly toward distillation units with multiple side draws. However, this chapter looks exclusively into the nuances of specialty chemical operations. In the domain of specialty chemicals, the requirements of a distillation column are somewhat different compared to traditional distillation operations (found in commodity chemical processes). The key features of these differences are described in the following sections.

8.1 Key Features

The key or principal feature of specialty chemical processes is the production of high product purities: usually >99.9%. These chemicals are used as feedstocks for the production of a variety of complex products that are very sensitive to feedstock impurities. For example, AA grade methanol is a specialty chemical that is used in the production of plastics and other compounds. Consequently, the purity of the AA grade methanol requires specific impurities, such as ethanol, to be reduced

A Real-time Approach to Distillation Process Control, First edition. Brent R. Young, Michael A. Taube, and Isuru A. Udugama.

Companion website: www.wiley.com/go/Young/DistillationProcessControl

down to parts per million (ppm) levels. Similar specifications can be found for other specialty chemicals. In addition, synthesis of most specialty chemicals, even with excellent catalyst selectivity, result in the production of multiple other compounds through side reactions. As a result, the incoming feed stream to the distillation train consists of multiple components, many in trace amounts.

In a traditional distillation process, it is common to "lump" these impurities into the heavy or light key compound both in the process design and controller design stage. However, with the need to manage product impurities at such low levels, there is a need to accurately understand the behavior of these impurities and devise control strategies that explicitly account for them. In particular, if they are "middle boiling" components (i.e. components that have a relative volatility between the main components in the bottom and distillate streams), then the level at which these "middle boiling" impurities present in the distillate (or bottoms) stream dictates the purity of that stream.

8.2 Measurement and Control Challenges

From a process control point of view, the requirement to determine product purities, where components present down to ppm levels, creates a unique set of challenges. In traditional bulk chemical distillation columns (as well as refining operations), the purity of the distillate or the bottoms stream is often determined by proxy. To be more specific, the temperature at the condenser or the reboiler, together with an understanding of the vapor–liquid equilibrium (VLE), is employed to determine the temperature at which the condenser or the reboiler must be kept to make a certain product purity. This works because any changes in the product purity will directly influence the VLE, which will then affect the temperature of the mixture. Thus, controlling the temperature at the reboiler and/or condenser means that one can control the purity of the product. However, this is not always the case for specialty chemical distillation processes for several reasons:

1. As the purity of the product decreases to ppm levels, even in a binary distillation process, the temperature at the condenser becomes less responsive to changes in the distillate purity. This is

because a change at ppm level does not significantly affect the VLE and, thus, the temperature becomes less significant as a proxy for composition. Consequently, any temperature changes will either not be detectable by the measurement device or be hidden within the measurement noise.

2. From a theoretical point of view, moving the temperature sensor to a tray below the condenser (or above the reboiler) might remedy this situation. For a given product concentration to be achieved at the distillate or bottoms streams, a certain composition must be achieved at a given tray (or point) in the distillation tower. Thus, a "proxy for the proxy" may be determined where a given tray temperature is maintained to make an on-specification product. While this type of configuration is often implemented in practice, significant operations effort is required to ensure that this measurement device is properly calibrated and maintained, so that the tray temperature "guarantees" on-specification production. It is worthwhile to note that, for grassroots projects, it is highly desirable to have extra nozzles installed on the tower which allow measurement from several different trays due to the uncertainties of tray efficiency and identification of the tray that reliably reflects the product composition. In brief, it is easier to have the tower built with extra nozzles (even if they end up not being needed) versus not having them, as retrofits can be prohibitively costly.
3. However, in specialty chemicals, where low or trace quantities of feed impurities are the norm, setting up a temperature-based control structure (as discussed in item number 2) may be influenced by "large" changes in the feed mixture, which may affect the VLE, and, thus, effect the proxy measurement. Hence, it is vital to perform a number of sensitivity studies on the tower design. The effect of feed composition variations must be well understood and the control structure(s) designed to account for it.

Given the potential for poor proxy measurements of product purity, installing a direct real-time online measurement of product composition using analytic techniques (such as gas chromatography or photometry) reduces the need for such proxies. However, this also requires special and dedicated maintenance efforts to ensure reliability and accuracy.

Similar to composition measurements (either inferential or direct), the requirement for high product purity also brings along other challenges. One such additional challenge is the potential for non-linear behavior. This simply refers to a given change in the manipulated variable in either size or direction that results in a disproportional response in the process variable of interest.

Figure 8.1 illustrates a hypothetical example where an impurity specification on a product is controlled by changing the distillation flow rate. In this instance, a nonlinearity can be clearly seen: Lowering the flow results in a smaller impact on the product impurity than does an increase in flow. It is also observed that this effect is further intensified when the magnitude of the change is made larger. In control terms, this represents a situation where the process gain (K_p) varies over regions of operation. Similarly, the time required to reach steady state (i.e. the process time constant) is also influenced and is different for each test.

Another aspect to this phenomenon was mentioned briefly in Chapter 5 regarding the potential for integrating behavior with composition measurements. The important factor to take away from this is that a "large" disturbance in mass balance will have a much greater proportional effect compared to a "small" change. Thus, it is vital to

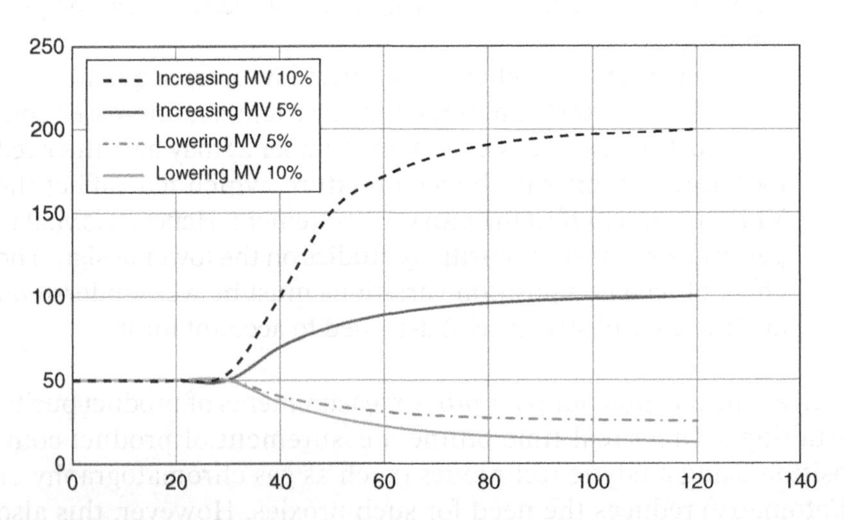

Figure 8.1 Product impurity specification versus distillation flow control. Where each equivalent control "move" results in a process gain and a time constant that are slightly different.

understand how trace, as well as main, components will accumulate in a distillation tower (i.e. how the composition profile will be established under design conditions) and the effect that changes in mass and energy balance have on it. The contributing factors to this issue are explored further in the sections that follow.

A remedy for many nonlinear behaviors is a concept called gain scheduling (e.g. Åström and Wittenmark 1984; Bett 2005; Ellis 2012). As discussed in the previous chapters, PID controller gain (K_c), integral time (T_i), and derivative time (T_d) are based on both steady-state and dynamic characteristics exhibited by the process. That is, based on assessing the process gain (K_p), the first-order process time constant (τ), as well as the process dead time (θ). However, in a situation where K_p, τ, and θ change significantly over the operating region, the controller tuning must also change based on the current operating point. This is achieved by implementing a precalculated "schedule," which automatically adjusts the PID controller tuning values as the process moves through different conditions. To implement any sort of characterization (i.e. gain or tuning scheduling), however, requires a very clear and precise understanding of how the process behavior changes and what condition(s) determines when it changes. As with many aspects of process control, making such assessments is not terribly difficult; it is just tedious and detailed!

8.3 Nuances of Fine Chemicals Distillation

Other than the nonlinearity issues and complexities in measurements, specialty distillation columns operations are also often affected by "large" time constants. As a result, there are two key issues that arise:

1. The existence of trace compounds in the feed, combined with the relatively large time constants, can result in the accumulation of these trace compounds over a "long" period, as alluded to in Section 8.2. As previously mentioned, compounds in the ppm range have little effect on the overall VLE. However, given the right conditions, trace compounds (especially if they are middle boiling compounds) can easily accumulate over time within a distillation column. This is similar to the accumulation of inert

compounds in recycling loops (e.g. Luyben et al. 1998; Svrcek et al. 2014). Thus, like in a recycle loop, these trace components must be purged from the distillation column. Failure to do so can easily result in a loss of separation and other unwanted behaviors. Thus, managing the overall component inventory (e.g. composition profile) must be performed as a part of normal process operations via appropriate control structures.

2. Managing the inventory of a trace component implicitly requires accurate measurements of flow and composition at all inlet and outlet streams, which is often impractical. Therefore, a good method to monitor the buildup of trace components is by following the overall temperature profile of the distillation column. This requires that a reasonable number of temperature measurements be installed throughout the column. The accumulation of a trace compound may, depending on the concentration, result in a distortion of the column temperature profile, as illustrated in Figure 8.2. This hypothetical example shows the accumulation of a trace component toward the condenser of the column. In this example, the feed to the column is at tray number 17 (numbered from the bottom). When compared to the normal temperature profile, one can see that the temperature profile at the top of the column first drops below the normal values between stages 19 and 29 and then goes above the normal values between stages 30 and 34.

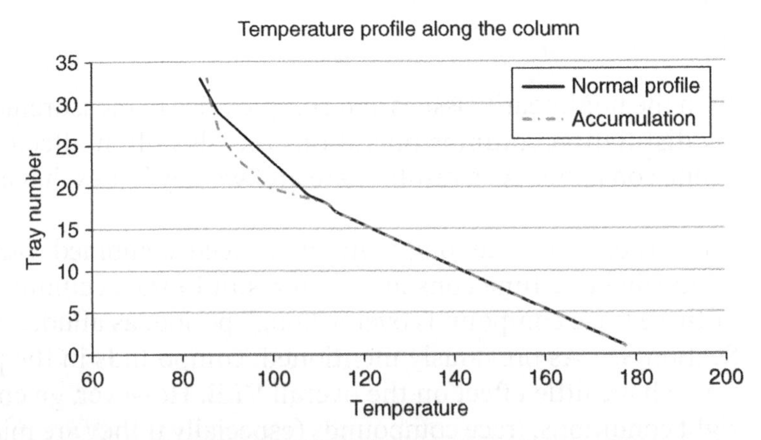

Figure 8.2 Temperature profile along the column with and without accumulation of a middle boiling component.

This type of distortion signifies the accumulation of a middle boiling trace compound. Once detected, necessary remedial operational action must be taken to periodically "flush" the accumulation or carry out design changes (such as introducing a side draw). While an experienced designer can identify the potential accumulation of a middle boiling trace compound at the design phase based on VLE data, precisely predicting the location of such accumulation is complicated by the uncertainties of tray/packing efficiency and the accuracy of the VLE properties. Hence many side-draw distillation columns will consist of multiple side-draw points, which can be used in operations.

The converse of component accumulation is the depletion of inventory. This is a specific issue in specialty distillation columns as high product purity (which, together with the commercial need to recover as much of the valuable products as possible) leads to a situation where all of the valuable compound (product) is channeled to the distillate (or bottoms) flow. Consequently, from a mass balance point of view, nearly 100% of the product component that is fed to the column is withdrawn from one stream. Thus, in a design where a traditional distillation composition control structure is employed to regulate the flow rate of product, the near 100% product recovery, along with inherent process nonlinearities and large time constants, means that slight variations in the process (which affect mass balance) can easily result in a situation where more than 100% of the product entering the column in the feed is removed from the column via the product stream. Consequently, there is a need to ensure the overall mass balance is continuously maintained.

One recommended solution is to configure a ratio controller that manipulates the distillate flow rate based on changes in the feed flow rate, as shown in Figure 8.3. While the product concentration in the feed is typically not measured "online" and flow measurements are typically not accurate (especially in turn-down scenarios), using a ratio control structure that is reset by an online product stream composition control will compensate for measurement errors, as shown in Figure 8.4. However, the more accurate the flow measurements, the closer the tower can be run to the specification limit. Many high-purity distillation columns are operated by "over refluxing,"

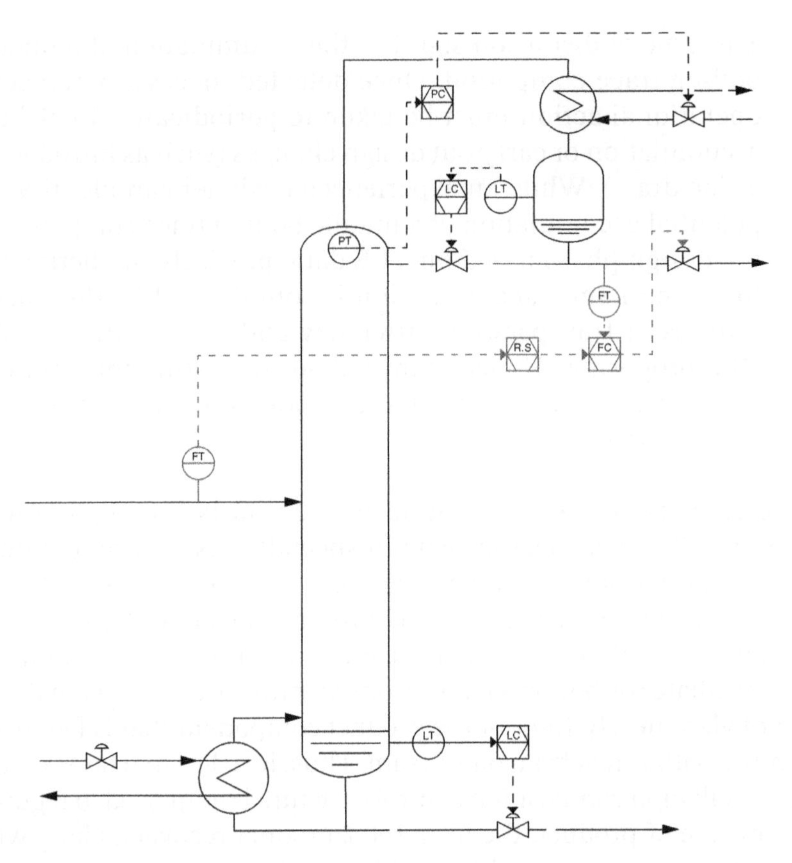

Figure 8.3 Ratio control configuration between feed flow and distillate flow to ensure the overall mass balance in maintained.

which refers to the practice of using a higher than required reflux ratio (i.e. energy input) to ensure the high-purity product specifications are met.

While this practice ensures (to some degree) on-specification production, it also reduces the potential maximum capacity of the tower, due to the higher vapor/liquid traffic (e.g. tray hydraulics). From an operator's perspective, (s)he desires to operate the tower a "comfortable" distance away from the specification limit. But this reduces the potential economic benefit. Thus, the proper design and tuning of the control structures is required to relieve the operator from "baby-sitting" the tower's operations, which is one of the purposes of this book!

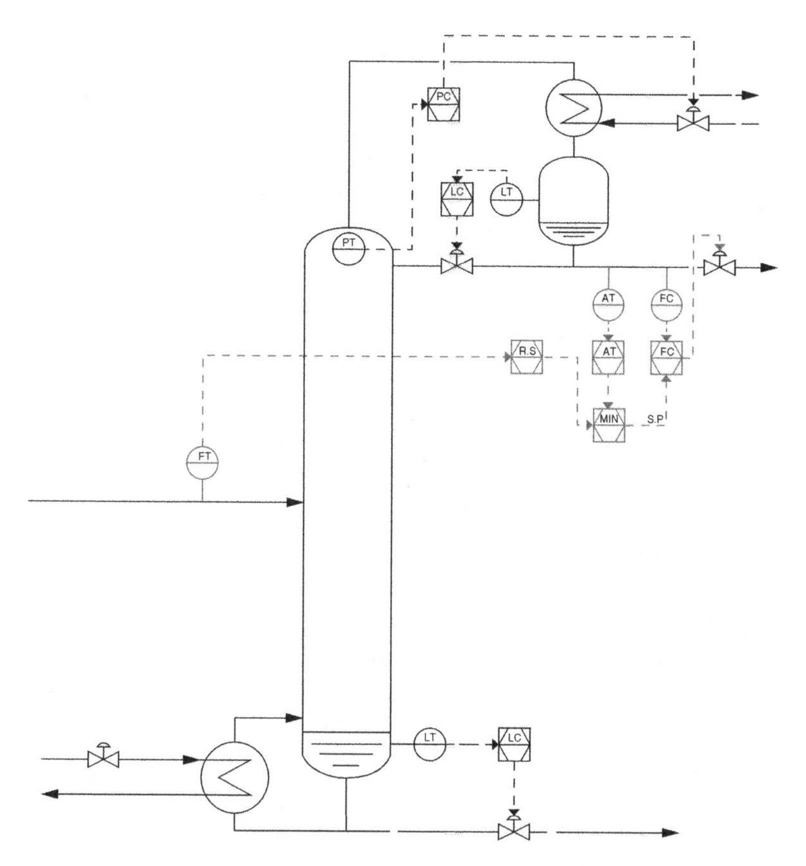

Figure 8.4 A ratio control structure reset by online product stream composition control.

8.4 Side-Draw Distillation

Implementing a side draw (e.g. a fusel oil draw in the distillation of alcohols) becomes necessary when the mixture refined to high purity contains trace components, which are middle boilers; as discussed in Section 8.3, these components will accumulate at different areas of the column over time. From a thermodynamic point of view, this accumulation happens because the middle boiling trace component has higher relative boiling point than the component(s) in the distillate, but at the same time it has a lower boiling point than the component(s) in the bottoms stream. As the description "middle boiler" asserts, the pure component boiling point lies between that of the "light" component(s) in the distillate and the "heavy" component(s) in the bottoms stream. If these impurities are not dealt with, they will continue to accumulate, which, over time, will influence both the overall VLE and the column gas and liquid loading. In short, this accumulation will result in a loss of separation (for the high purity product) even at elevated reflux ratios. Operationally, the only option to remedy such a situation is to "flush out" the trace impurities, which requires that production be halted. A more practical and efficient approach is to introduce a side draw. The key benefit of side-draw distillation columns is their ability to perform two separation operations in one column and, thus, they are often considered a form of process intensification. As with many intensified operations, the process design is a trade-off between process controllability and improved process efficiency. The side-draw distillation clearly illustrates this trade-off.

Before defining the design criteria for composition control, one must first establish the base flow and inventory control requirements for side-draw towers.

In a situation where the side-draw component consists of a large percentage of the feed stream, the control structure shown in Figure 8.6 is recommended. This is because the side-draw rate is larger than the bottoms draw rate. As such, the bottoms draw would have a potentially smaller influence on the level (i.e. the bottoms stream has a smaller relative gain versus the side stream). Nevertheless, the tower's mechanical design of the bottom's accumulator will affect this choice.

Conversely, if the side stream flow, as well as the Heavies steam, represents only a small fraction of the feed, then an alternative control structure is warranted: most likely one where the side stream is the MV for a temperature controller for the tower.

8.5 Composition Control in High-Purity Side-Draw Distillation

Due to all of the possible combinations of feed compositions, product quality control in a high-purity distillation column with a side draw is challenging. Thus, any attempts to propose a "cookie-cutter" solution for this type of column are impractical. The first distinction that needs to be made is whether the side draw requires explicit control or

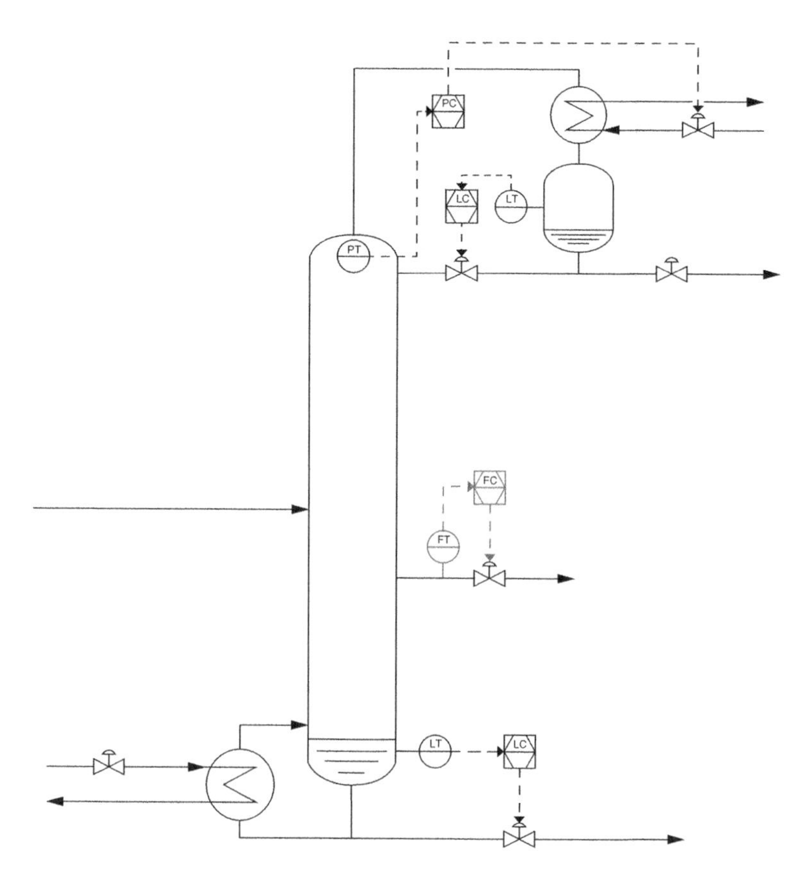

Figure 8.5 Side-draw flow controller arrangement for a column configuration where the side-draw flow is low.

the high-purity control loops (illustrated in Figures 8.5 and 8.6) are sufficient to maintain on-specification production. The answer to this question lies with the feed stream, that is:

1. If the feed flow rate and the composition of the "middle boiling" components in the feed that must be removed through the side draw remain at a steady level over a long period, then only passive monitoring and occasional adjustment of the side-draw flow rate set point to ensure no accumulation of these components.
2. However, suppose the feed flow rate and the composition of the middle boiling components in the feed constitutes more than trace levels and/or varies considerably. In that case, active control measures must be used.

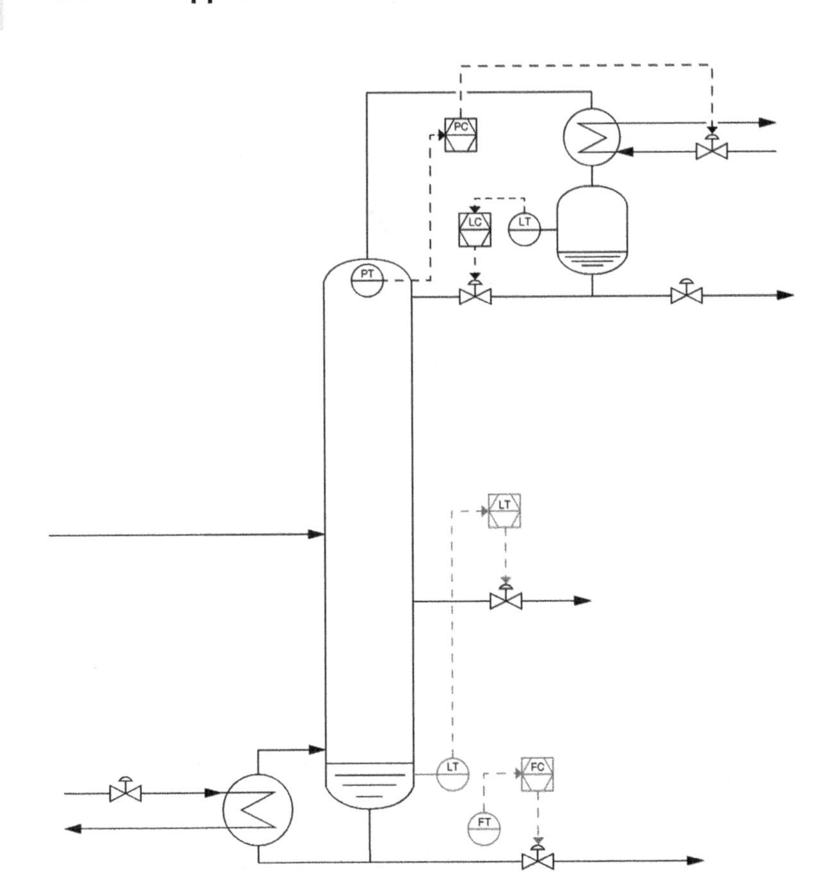

Figure 8.6 Level control arrangement using the side-draw controller for a column configuration where the side-draw flow is larger than the bottoms flow.

One such industrial example of the second case is industrial methanol production, where a multicomponent feed of methanol, water, and ppm levels of ethanol is fractionated in a distillation column with a side draw. In addition to the very stringent product and bottoms purity specifications (ppm levels), it is also required that methanol recovery exceed 97% (as can be seen from the specifications in Table 8.1).

From a commercial point of view, the distillate methanol needs to contain <10 ppm ethanol to meet the Federal AA grade specification, while a high recovery rate (97.5%) ensures high profitability. Almost all the water entering the column is withdrawn in the bottoms stream, where an average methanol composition of <10 ppm must be maintained to ensure that this stream can be discharged without the need for costly bio-treatment. These restrictions create a unique

Table 8.1 Example specifications for a methanol distillation column with a fusel oil side draw.

Stream	Quantity	Plant data	Units
Feed	Flow	14200	kg/h
	Methanol	0.826	Mass frac (%)
	Ethanol	150	ppm
	Water	0.179	Mass frac (%)
	Butanol	95	ppm
Distillate	Flow	114500	kg/h
	Methanol	99.99	Mass frac (%)
	Ethanol	7	ppm
	Water	85	ppm
	Butanol	0	ppm
Bottom	Flow	24000	kg/h
	Methanol	7	ppm
	Ethanol	7	ppm
	Water	99.94	Mass frac (%)
	Butanol	0.05	Mass frac (%)
Fusel	Flow	3100	kg/h
	Methanol	85.54	Mass frac (%)
	Ethanol	0.66	Mass frac (%)
	Water	13.8	Mass frac (%)
	Butanol	47	ppm

ethanol profile along the column, resulting in an ethanol "bulge" near the tower bottoms. The formation of this bulge allows for 95% of the ethanol entering the column to be removed via a side draw. In comparison to typical binary distillation columns, the control of an industrial methanol distillation column requires a control scheme that accounts for:

- A multicomponent feed stream
- High recovery rates (97.5%)

- The management of a non-key, mission-critical ethanol product composition
- The management of the ethanol composition at the side draw

Process control structures must be implemented to detect the composition bulge's movement and effectively control the bulge (e.g. using reboiler duty to maintain bulge shape and position). Alternatively, tight overall control can be implemented to ensure the ethanol bulge is passively controlled while actively controlling the side-draw flow rate to counter feed flow and composition variations.

8.6 Advanced Distillation Column Configurations

The performance and/or efficiency of side-draw distillation columns can be further improved through mass and/or heat integration. The general objective is to achieve greater purity of the distillate, bottoms, and side-draw streams. In pursuing these improvements, the utilization of more complex column configurations is explored.

One modification that may be carried out is the addition of a pre-fractionation column (or flash drum) to the main column (Figure 8.7). In this case, the pre-fractionation column is used to make an initial "split" in the feed into vapor and liquid, which is then introduced at two different locations in the main column. In this instance, the vapor and liquid splits are a function of the process feed composition and enthalpy, and no "control" can be carried out on these streams (other than liquid inventory). In addition, the pressure at the prefractionator is also not actively controlled. Rather, the overall pressure profile of the main column, as well as the feed flow pressure, dictates the pressure profile in the prefractionator section. Similarly, the thermodynamics and feed composition dictate the temperature (composition) inside this vessel. However, the pressure and temperature in the pre-fractionator can be actively monitored and influenced by the reboiler and condenser duty of the main column.

8.7 Petlyuk and Divided Wall Columns

The pre-fractionation design can be further improved by implementing a Petlyuk configuration (e.g. Petlyuk et al. 1965; Halvorsen and Skogestad 1999). This results in a mass-integrated design where flows

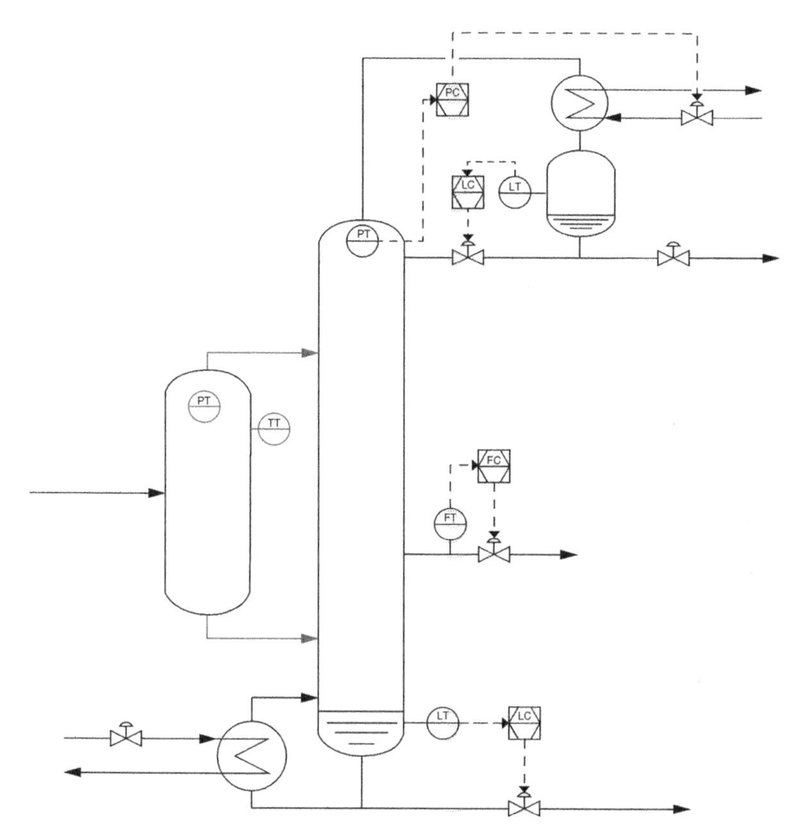

Figure 8.7 An intensified column design consisting of a pre-fractionation unit.

are exchanged between the pre-fractionation and the main column sections, enabling improved process efficiencies. Despite the noticeable "exchange" of flows between the main column and the pre-fractionator, only a single meaningful control valve can be installed in such a column: the liquid split between the main column and top of the pre-fractionator is illustrated in Figure 8.8. In contrast, the other mass flows cannot be controlled due to the following:

1. Vapor flow from the pre-fractionator is based solely on the feed stream's enthalpy and the pressure drop as it enters the vessel. Thus, the vapor fraction leaving the vessel to the main tower cannot be "controlled" independently of the thermodynamic relationships.
2. The same is true for the liquid leaving the vessel. While in practice, there often is a liquid inventory maintained in these

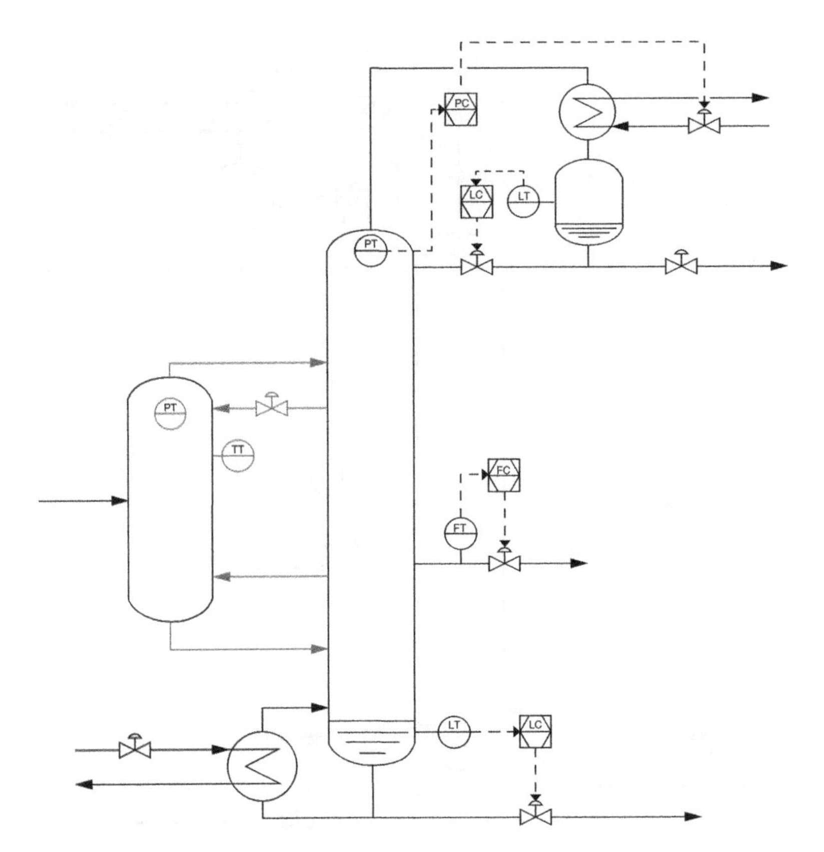

Figure 8.8 Potential control handles that can be used to control a pre-fractionation unit.

types of vessels, including a level controller, it is intended to ensure that no net accumulation occurs due to changing feed conditions.

Coupling heat integration to the Petlyuk design yields an even further integrated column concept known as a divided wall column (e.g. Lorenz et al. 2018) as illustrated in Figure 8.9. In a divided wall column, the pre-fractionation is physically integrated into the main column and the two separate vessels are replaced by a single divided wall column. To facilitate the liquid distribution at the top of the distillation column, a dedicated liquid distributor is used, which, like in the pre-fractionation column (Figure 8.8), is intended to provide some degree of control in the liquid split to either side of the wall. This is the only "independent variable" that can have some effect on the process. The liquid split at

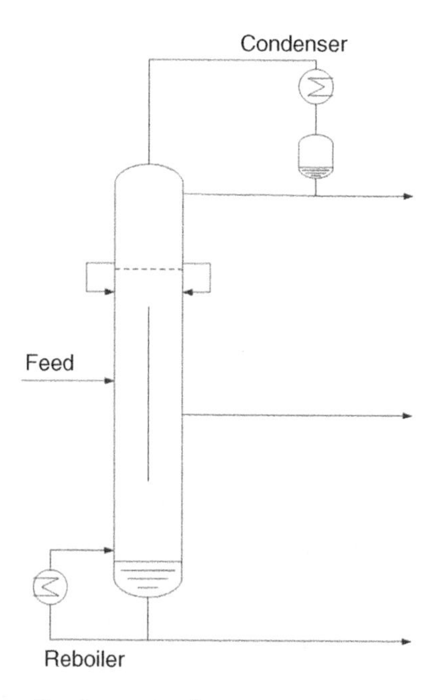

Figure 8.9 A divided wall column configuration.

the bottom, as well as the vapor flow rates into and out of the divided wall, cannot be controlled for the same reasons discussed in the pre-fractionation example.

From a practical point of view, the liquid split at the top of the pre-fractionator/divided wall must be manipulated to maintain the liquid/vapor loading pre-determined by the design of the column such that the column vapor liquid loadings are satisfied. Hence, manipulating this distribution cannot be treated as an independent variable for control purposes (much like reflux in a typical binary distillation tower).

Based on the above, a key conclusion about these intensified distillation column configurations is that, despite process integration, the number of independent variables needed to control the process is practically zero. Furthermore, compared to a set of binary distillation columns performing a similar separation process, each of these intensified column designs (including the side draw column) has lost some degree of controllability due to the loss of degrees of freedom. Thus, it is vital that the mechanical design of such intensified processes must ensure that the necessary degrees of freedom are provided.

8.8 Optimal Design Versus Optimal Operations

When a new unit design is being developed, the process designers often attempt to come up with an “optimal solution” around the process’ design point. However, once the plant is built and is operational, rarely, if ever, it is operating AT the design point! To make matters worse, process control input for the unit is usually engaged well AFTER the mechanical design is complete and fabrication well underway. Thus, any considerations for stability, turn-down, and/or maximum throughput that could be addressed mechanically are not viable (without adversely affecting the project schedule and budget). Hence, a key factor that must be considered when analyzing complex column configurations is the controllability of the column(s) throughout its (or their) entire operating envelope(s). In general, increasing the heat and mass integratation of a column design reduces its flexibility while simultaneously increasing the complexity of the controls required to achieve stable operations. Therefore, it is imperative that all new unit designs consider control ability and stability across the operating spectrum, rather than as an afterthought.

In short, process control and instrumentation expertise must be engaged much earlier in a new unit’s design. Otherwise, unexpected operating limitations are bound to occur.

8.9 Conclusions

This chapter focused on distilling specialty chemicals where high product purities must be obtained from a multicomponent feedstock. As a result of the feed specifications and product requirements, the fine chemical distillation processes often can benefit from intensified distillation column configurations such as divided wall columns. While these intensified distillation configurations may improve aspects such as energy usage or product recovery specification, the control of these intensified configurations requires more complex control arrangements.

Tutorial and Self-Study Questions

8.1 How is nonlinear behavior most easily identified? How can/would one address it? Where is it most often found? Why is it important to make processes appear linear?

8.2 Why is it better to use direct composition analysis in high-purity distillation operations?

8.3 What are the potential operational issues that can arise from intensified distillation operations (lost DOF)?

8.4 What should you consider when contemplating advanced design configurations?

8.5 What are the trade-offs of implementing a side-draw distillation configuration?

8.6 What is the best way to control a composition bulge in a side-draw distillation column?

References

Åström, K.J. and Wittenmark, B., (1984). *Computer Controlled Systems: Theory and Design*, Prentice-Hall International, 351–352.

Bett, C.J. (2005). Gain-scheduled controllers. In: *The Electrical Engineering Handbook* (ed. W.-K. Chen), 1107–1114. Academic Press.

Ellis, G. (2012). Nonlinear behavior and time variation. In: *Control System Design Guide*, 4e (ed. G. Ellis), 235–260. Butterworth-Heinemann.

Halvorsen, I.J. and Skogestad, S. (1999). Optimal operation of Petlyuk distillation: steady-state behavior. *Journal of Process Control* 9 (5): 407–424.

Lorenz, H.-M., Staak, D., Grützner, T., and Repke, J.-U. (2018). Divided wall columns: usefulness and challenges. *Chemical Engineering Transactions* 69: 229–234.

Luyben, W.L., Tyreus, B.D., and Luyben, M.L. (1998). *Plant Wide Process Control.* New York: McGraw-Hill.

Petlyuk, F.B., Platonov, V.M., and Slavinskii, D.M. (1965). Thermodynamically optimal method for separating multicomponent mixtures. *Khim Engng* 5: 555.

Svrcek, W.Y., Mahoney, D.P., and Young, B.R. (2014). *A Real-Time Approach to Process Control*, 3e, 236–241. Wiley.

9 Advanced Regulatory Control

9.1 Introduction

Advanced regulatory control (ARC), also referred to as supervisory control, encompasses many complex distillation column control configurations that strive to go beyond achieving "stability": ARC is the first step in moving a control strategy from maintaining stable operations to achieving smoother operations in the face of measured disturbances, which then leads to improved overall economics. However, while ARC does not explicitly "optimize" operations, it provides a foundation upon which model-predictive control (MPC, addressed in Chapter 10) and/or real-time optimization (RTO) can then optimize a plant or unit's performance.

Like a foundation under a building, MPC/RTO relies on a solid foundation of properly structured controls and well-maintained instrumentation. In the case of distillation, the foundation for all control endeavors is the base inventory and composition control that were covered in the previous chapters. As illustrated in Figure 9.1, the ARC acts through this regulatory layer foundation, while the MPC/RTO layers act through the regulatory and the ARC layer. This hierarchical approach to control allows the higher-level layers to focus on executing the slower but computationally increasing complex tasks while relying on the lower layers to carry out higher frequency and simpler control operations. It should also be noted that the top three layers of the control structure hierarchy focus on "plant-wide" aspects of operations and not the individual unit operations. The ramifications of the decisions made in these top three layers of distillation operations will be discussed in Chapter 11.

A Real-time Approach to Distillation Process Control, First edition. Brent R. Young, Michael A. Taube, and Isuru A. Udugama.

Companion website: www.wiley.com/go/Young/DistillationProcessControl

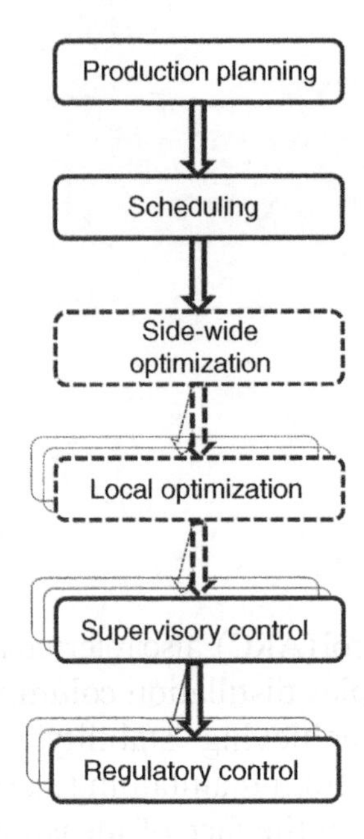

Figure 9.1 Process control architecture layers.

In an ideal world, the ARC control layer can be developed on top of the base control concepts covered in the previous chapters (including composition control), without re-visiting the regulatory layer design. However, in practice, the control loop where ARC control is introduced must be reassessed and if necessary restructured. The non-affected control loops may be kept the same as previously designed as long as these loops are not affected by implementing the ARC control layer.

9.2 Cascade Control

Cascade control (e.g. Murril 1967; Shinskey 1967; Svrcek et al. 2014) is a standard ARC technique that is implemented to capture and react to measured process disturbances that can affect the performance of the controlled variable (CV) of a regulatory control loop. Thus, cascade control improves the “speed” at which a control scheme reacts to measured disturbances in the process that would ultimately affect the key process variable of interest.

Cascade control achieves this by using two nested loops (a loop within a loop). The outer loop in this configuration will monitor the key process variable of interest that must be kept on target. While the inner loop is controlling another related process variable which, if not managed, will affect the key process variable. Therefore, if a disturbance is influencing the process, the inner loop will detect this variation and act to alter the final control element (typically a valve) before the disturbance affects the key variable, while, for any other disturbance affecting the key variable, the outer loop will "pick up" the variation in its CV and act by changing the set point of the inner loop (which, in turn, will result in the inner loop acting on the final control element). To ensure the correct operation of this type of control structure, the inner loop must be acting on a "faster" CV than the outer loop. That is, the behavior of the inner loop's CV–MV relationship is much faster dynamically than the outer loop's. This also ensures that the inner loop can keep up with the set-point changes that are issued by the outer loop. Typically, a 4 : 1 ratio of out-to-inner loop time constants is expected, which translates to four times lower time constant for the inner loop than the outer loop.

9.2.1 Cascade Control in Distillation

Condenser cooling water/cooling air and reboiler steam/fluid flow are two main areas where a cascade control is beneficial in achieving improved operations in a distillation process. This is because the condenser cooling water/cooling air is typically used to control the overall column pressure, which is important to ensure suitable conditions to carry out a separation process, while the reboiler steam/fluid flow provides the thermodynamic "driving force" to enable the separation process to take place. To this end, any unintended variations in these critical flow rates can have significant operational ramifications on a distillation column. Moreover, both the cooling and heat flows are typically supplied to the distillation unit by plant utilities, which are heat integrated to other parts of the plant. For example, in many chemical facilities, the steam used in a distillation operation is created by cooling down an exothermic unit operation (e.g. a reactor). As a result, these flows can easily have significant variations, particularly the pressure at which they are supplied to the distillation unit. To counter these variations, the following cascade control structures are recommended (Figure 9.2).

The control structure in Figure 9.3 (the control loops in black) is based on the previous double ended composition control structure

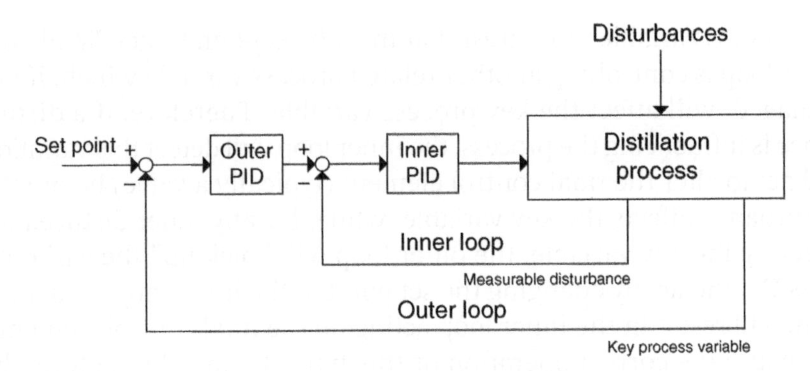

Figure 9.2 Cascade control general information flow and components.

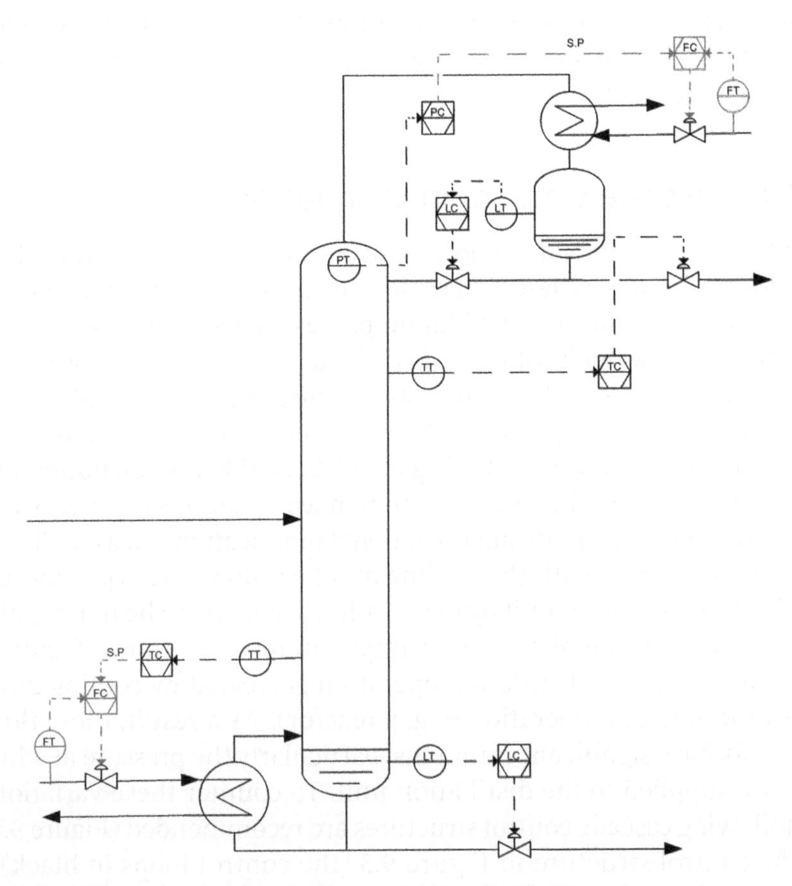

Figure 9.3 An example of cascade control in a distillation column.

discussed in Chapter 5. While the control loops in red illustrate the alterations made. Starting off with the reboiler controller, rather than directly manipulating the steam (hot liquid) flow into the reboiler, an inner flow control loop is introduced where the flow rate of stream into the reboiler is monitored and controlled. The set point of this flow control loop is determined by the outer temperature control loop that is maintaining a desired temperature control set point. Similarly, the condenser cooling water (fluid) flow is monitored and controlled by a flow control loop that gets its set point from the outer pressure control loop that maintains a desired pressure in the column.

In both these instances, any variation in the hot or cold fluid flows will be picked up by the flow transmitter and corrected through the flow control loop before it affects the bottoms temperature or column pressure, respectively. To this end, cascade control allows for potential disturbances in the utilities to be corrected for prior to them influencing the distillation unit operations. However, in both these instances, any change in enthalpy of the streams (energy/unit) will not be picked up by the flow control loops, as such these un-measured (as well as unknown) disturbances will propagate to the Distillation process, where they will be rejected from the outer control loop. To this end, the inner loop in a cascade controller will only correct for measured disturbances. However, if appropriate measurements are included (e.g. heating or cooling medium pressures and temperature), then another inner loop should be implemented to adjust the fluid flow before the enthalpy disturbance affects the key CV. In practice, such a controller is called a duty controller since its function is to maintain a specific thermal input or removal on the tower.

9.2.2 Inferential Cascade Control

When analyzing the function of a generic cascade controller, it should be observed that a side benefit of monitoring a disturbance variable is the ability to detect and react to known process disturbances in a rapid way. Inferential cascade control is a specific subset of cascade control that benefits from this rapid detection. More specifically, in an inferential cascade, the inner loop gets a process variable (measurement) that has no measurement delay but might suffer from a lack of precision or robustness (e.g. sensor drift). In contrast, the outer loop gets a process variable (measurement) that is accurate and robust but has some measurable time delay. Together, these two loops allow for a manipulated variable to react rapidly to disturbances in the process (picked up by the inner loop) while still ensuring on-specification operations with no drift over time through the outer loop. A well-documented example of such an inferential cascade system is the composition/temperature inferential used to maintain the on-specific composition of a key product stream such as the distillate stream, as shown in Figure 9.4.

In this case, temperature (which is an inferential for composition) is used in combination with an online analyzer (such as a gas chromatograph). The inner (temperature) control loop will maintain the desired temperature set point in the column while reacting to any process disturbances, whereas the outer (online analyzer) control loop will manipulate the set point of the inner loop when required to maintain on specification production. In principle, as long as the composition of the feed stream is constant and no other significant changes occur in the other parts of the distillation column, the inner loop should be able to maintain the on-specification production even when hit by process disturbances. However, if the composition of the feed stream

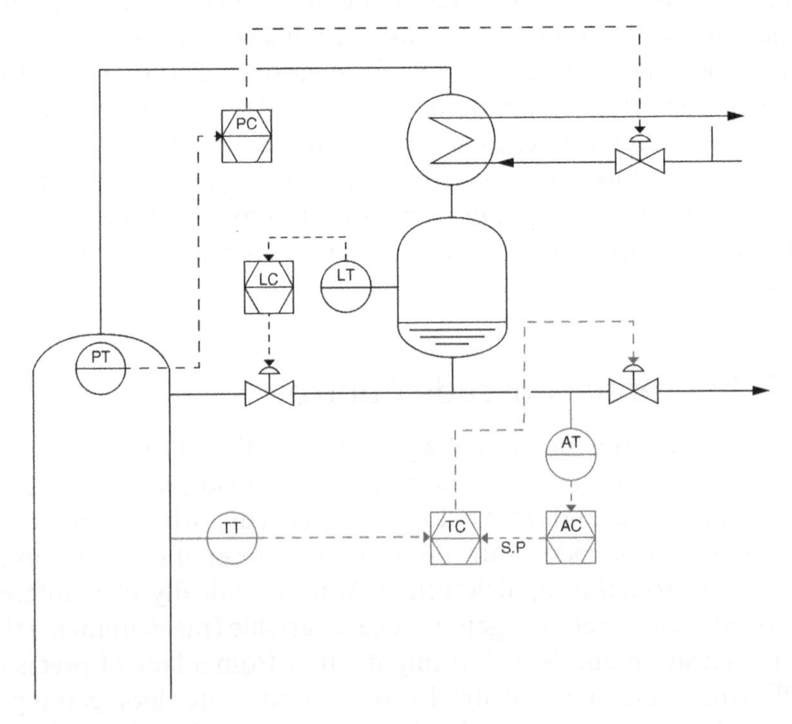

Figure 9.4 An inferential cascade control example.

changes and/or there are other changes in the unit's operations, then these variations may not be readily detected by the inner (temperature) control loop. To this end, the outer (online analyzer) loop is needed to detect the effect upon composition caused by these variations and update the set point of the inner temperature control loop.

There is, however, an unintended consequence to this kind of structure: "double-dipping." That is a disturbance that effects the temperature will, eventually, manifest in the online composition measurement. This, in turn, will cause the composition controller to *inappropriately* adjust the temperature controller's SP, since the temperature controller has already compensated for the disturbance. While there is a solution to this dilemma, it is outside the scope of this book.

9.3 Ratio Control

Ratio control is an ARC technique that is widely used to maintain critical ratios between process variables in process operations (e.g. Eckman 1945; Murril 1967; Shinskey 1967; Svrcek et al. 2014). In general, ratio control adjusts a process flow that must be maintained at a given ratio with another process flow that is prone to variation. Figure 9.5 illustrates a ratio controller's general information flow and components.

While there are several possible configurations by which a ratio control scheme can be implemented, the configuration above is recommended as it clearly conveys the nature of the control structure. In this

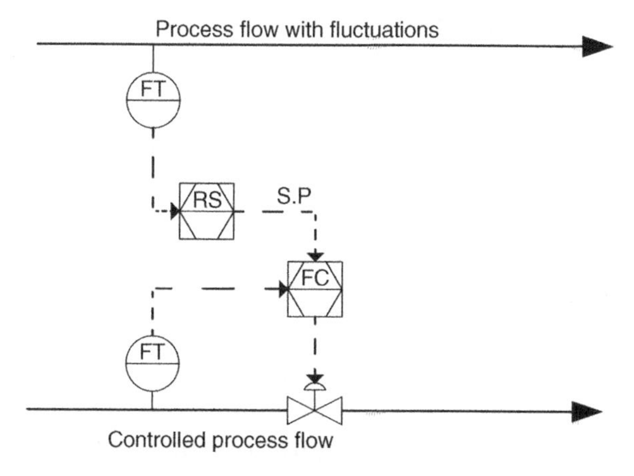

Figure 9.5 Ratio control general information flow and components.

scheme, a flow transmitter is used to capture the fluctuations in the uncontrolled/varying process flow; this value is then simply multiplied in a ratio station (RS) based on a given set point (such as a stoichiometric coefficient). The output of the RS is then used as the set point for a flow control loop that controls the process flow rate such that the ratio between the fluctuating, or "wild," flow and the controlled flow is always maintained.

Note that, in this context, the "wild flow" is often set by another regulatory control structure, such as an upstream inventory controller. But, from the perspective of the target flow (i.e. the one being manipulated by the RS), it is "wild" or uncontrolled. Also, it must be pointed out that the RS, often labeled as a ratio controller, does *NOT* perform any "control," it merely performs a simple mathematical operation: multiplying the ratio SP by the value of the "wild flow." Lastly, in practice, it is often required to perform some "dynamic filtering" of the wild flow (before it is multiplied by the ratio SP) in order to have a coordinated effect on another process measurement, such as a distillation try/stage temperature. This is touched on in Section 9.4.

9.3.1 Ratio Control in Distillation

In distillation operations, the ratio between mass and energy flow rates is key to achieving stable, on-specification operations. The following examples are often found in academic treatments, but rarely found in industrial implementations.

9.3.1.1 Reflux Ratio Control

One of the key ratios in a distillation operation is the reflux to distillate ratio.

As discussed in Chapter 2, the reflux of the distillate flow must be committed to control the inventory in the reflux drum, while the other flow rate is employed to control the temperature in the top portion of the column above the feed. With a traditional composition control configuration, this means any variations in the distillate stream will result in a change to the reflux flowrate. This, in turn, changes the reflux ratio (ratio of reflux flow to distillate flow), resulting in a shift in the column's operating line that manifests itself as a fluctuation in column temperature. A reflux ratio control configuration such as the configuration as illustrated below in Figure 9.6 can better counter these variations.

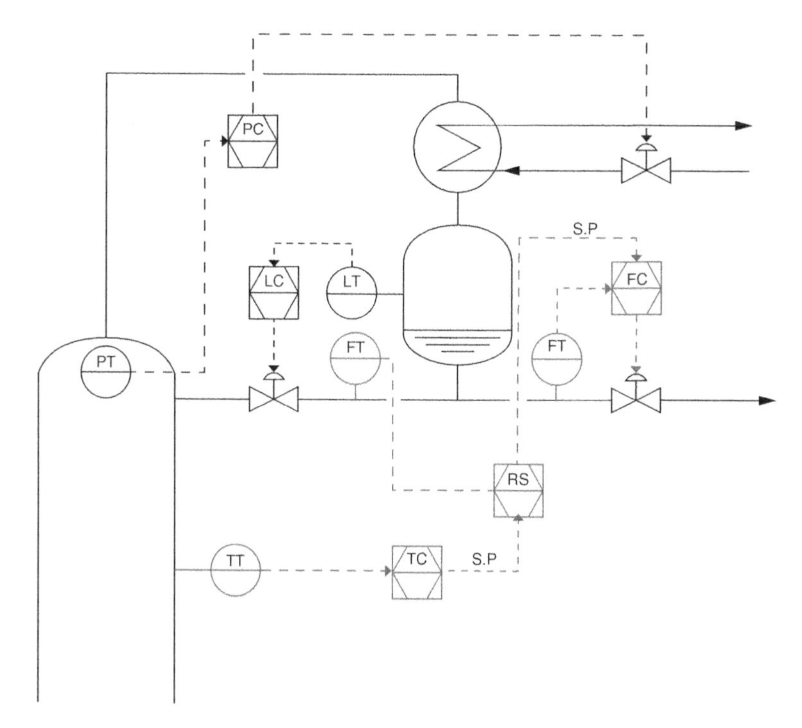

Figure 9.6 An example of ratio control in a distillation column.

However, in practice, it has been found that well-tuned inventory and composition controls effectively handle disturbances and do not require the additional complexity of this well-intended configuration.

9.3.1.2 Double Ratio Control

The concept of ratio control also applies to the boil-up rate at the bottom of the column where a similar control configuration as the reflux ratio control can be implemented. The ratio between the bottoms flow and the reboiler boil-up flow is maintained in this case. A double ratio control is shown in Figure 9.7. Again, such control structures are typically found only in academic settings and, rarely, if ever, in industry.

9.4 Feedforward Control

Feedforward control (e.g. Murril 1967; Shinskey 1967; Svrcek et al. 2014) is a higher order ARC solution and is the first step toward model-based control decision-making. Feedforward control attempts

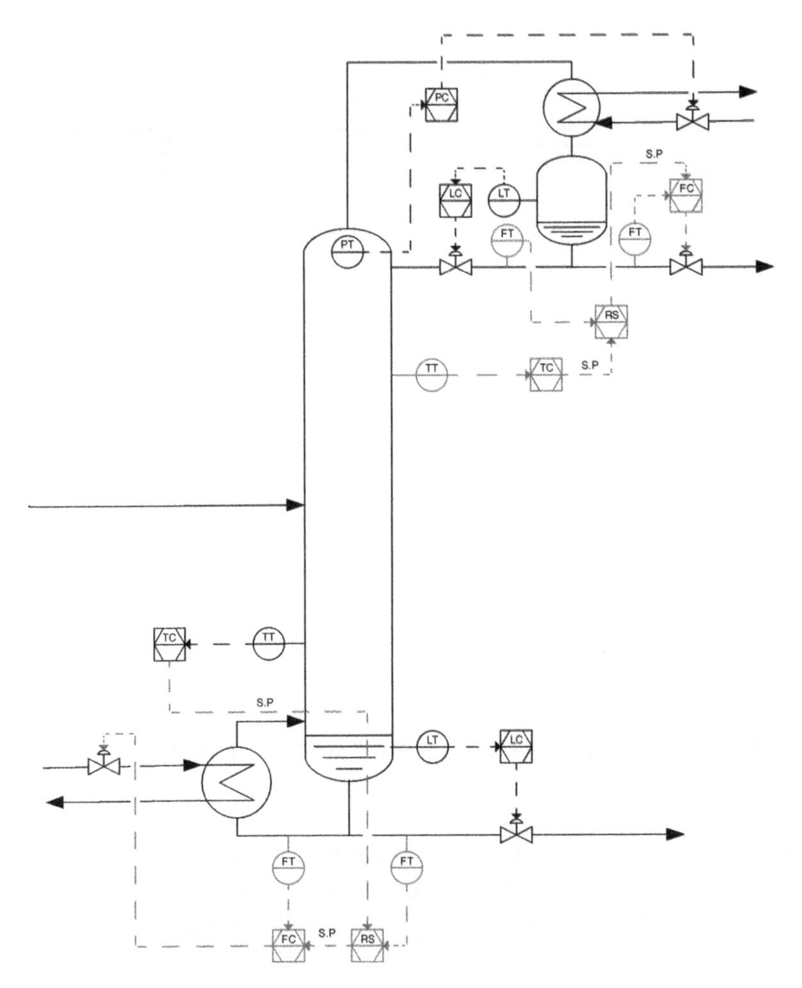

Figure 9.7 An example of double ratio control in distillation.

to pre-empt disturbances from propagating through a unit operation by detecting a process disturbance at the source and carrying out the necessary control actions to mitigate the influence of this disturbance, as illustrated in Figure 9.8. In simple terms, the feedforward takes control actions that attempt to cancel out the process disturbances' effect on the unit operation. To do so, a feedforward controller must be designed based on how the process will react in the "future" to a given process disturbance. To this end, a prerequisite for applying feedforward control is the ability to measure a process disturbance. Similarly, an accurate process model that relates current process disturbances with "future" process behavior is needed such that necessary

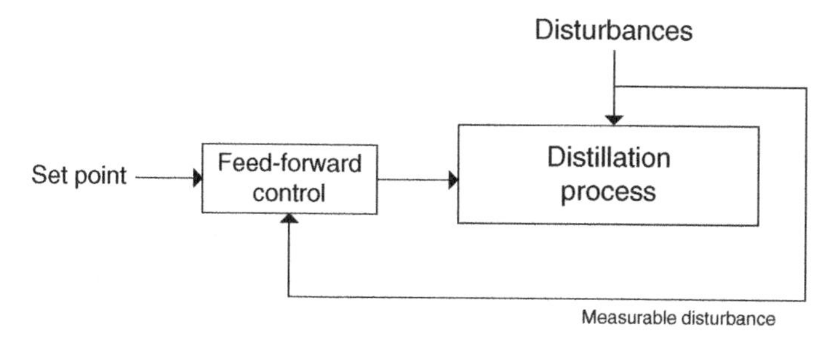

Figure 9.8 Feedforward control general information flow and components.

pre-emptive corrective actions can be taken to negate the effect of the disturbance.

A simple form of feedforward control includes setting the reboiler duty or a distillate flow rate for a given feed rate. The more advanced and widely implemented form of a feedforward controller includes a dynamic lead-lag filter, where the dynamics of how the disturbance(s) affects the process are taken into account in the feedforward control action. If a RS is included in the control structure, then the steady-state relationship is addressed implicitly. If not, then the steady-state relationship between the "wild flow" and the RS's MV must be explicitly included.

In practice, a feedforward controller in distillation operations consists of both a feedforward part, as well as a feedback controller. This is to ensure that any mismatches in the process model used in the feedforward controller are corrected by the feedback control structure. The primary objective of a feedforward controller is to negate the impact of a measured disturbance on a distillation operation by taking "pre-emptive" actions. Some illustrative examples follow.

The first case illustrated is a feed-to-reboiler duty control, shown in Figure 9.9. In this instance, a feed disturbance propagating through the plant can easily create a significant disturbance in the distillation unit. Thus, keeping the ratio of energy (reboiler duty) per mass flow entering the column (flow rate) will allow for smooth process operations: the lead-lag takes into account the dynamic effect of the disturbance (the feed flow) and the manipulated variable (reboiler duty) on the overall process such that they are coordinated to perfectly cancel the disturbances' effect. In this instance, the feedback controller is a simple temperature control loop, which will ensure the bottoms temperature is maintained at the required set point and negate any model mismatches. This is a common implementation found in industrial installations.

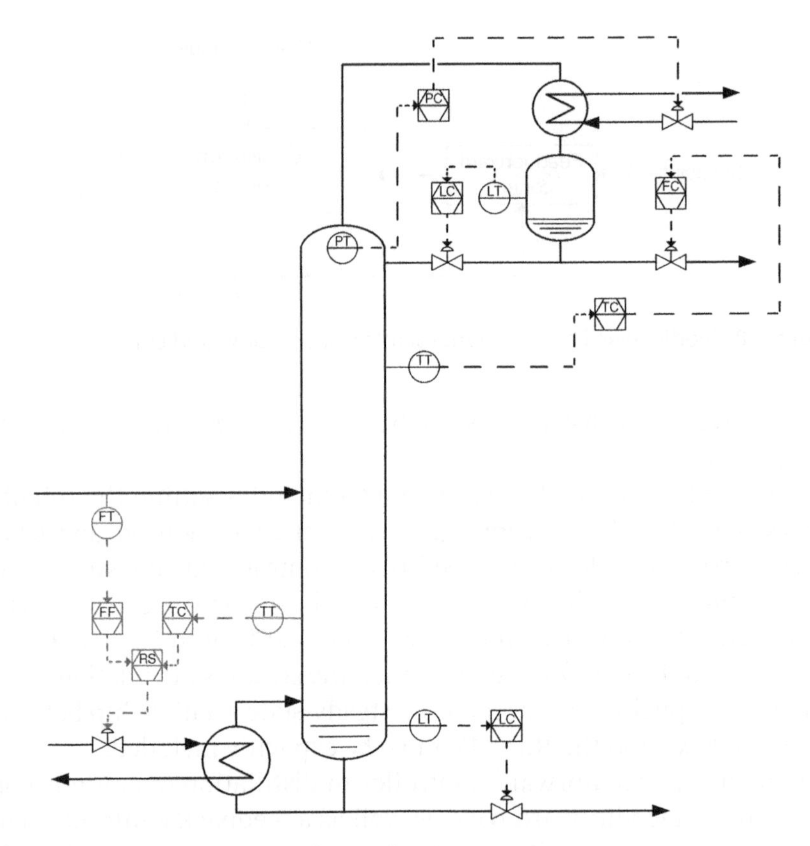

Figure 9.9 An example of feedforward ratio control.

9.5 Constraint/Override Control

A key factor that must be taken into account by process control engineers are the physical limitations of the process equipment for which they are designing the control structures. In the modern petrochemical industry, process designs are "squeezed" to be just capable of meeting a given production target and specification. This is often referred to as "value engineering." Value engineering, to some degree, was a reaction to the significant over-design practices of decades past where a plant's rated capacity was easily increased by 50–100% and higher than the original design. Now, however, the function of modern process control engineers is more challenging as there are significant equipment constraints (hard constraints that cannot be exceeded) and operating constraints (soft targets that must be satisfied) that greatly narrows the operating envelope. These kinds of issues are classically handled by designing the optimal solution into the ARCs (e.g. Shinskey 1967; Seborg 1984; Svrcek et al. 2014).

An example of such an optimal design solution is explained by further analyzing the previous control structure of feed-to-reboiler feedforward control. In this instance, if the feed flow rate is increased, the feedforward controller would also adjust the reboiler duty. As these actions are taken, the column's internal vapor and liquid traffic will increase: even if the reflux-to-distillate ratio is constant, there is now more mass entering the column and the increased reboiler duty results in increased vapor generation. The increase in vapor–liquid traffic in the column, if allowed to increase unabated, ultimately results in flooding. One way to tackle this type of issue is to implement differential pressure monitoring.

To control such a situation, the only meaningful control handle (that ensure some on-specification production is carried out) is to limit the reboiler duty based on the differential pressure across the tower (e.g. a measurement of tower flooding), as shown in Figure 9.10. As shown in the figure, this type of control structure uses a low selector function block: it passes on the lower of the two outputs: one from the Delta-P Controller and the other from the temperature/ratio controller. This is a very powerful tool for providing a "back stop" or protection for controller action that would otherwise result in an upset of other unwanted outcome.

9.6 Decoupling

Another common issue faced in distillation control is loop interaction (e.g. Seborg 1984; Svrcek et al. 2014). In particular, this can be seen in double-ended composition control, where a control configuration like LV (temperature and the top and the bottom of the column are controlled by reflux flow and reboiler duty) can lead to significant loop interaction. In fact, due to the intricate relationship between mass and energy balance in a distillation column, many control structures that attempt to control of the columns fractionation and product specification would have some degree of loop interaction. However, in most cases these loop interactions are not significantly detrimental to the

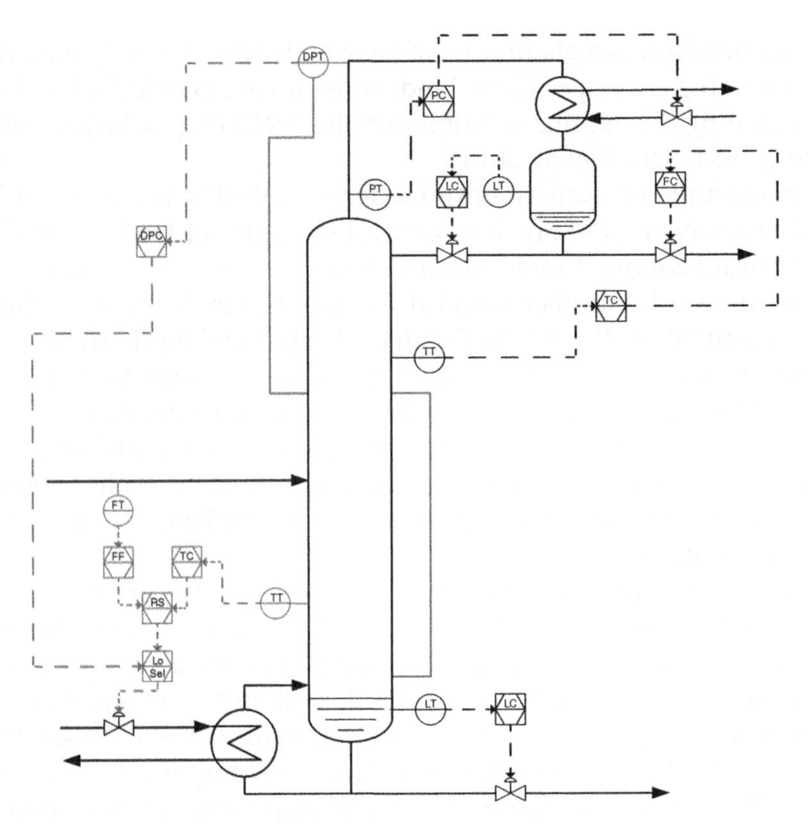

Figure 9.10 Override control to prevent tower flooding.

controller performance and hence further improvements to the control structure are not warranted.

At the same time, some control configurations will inherently have loop interactions that are detrimental to the overall controller performance. The simplest possible way to deal with this is to change the control structure so that loop interactions are eliminated or minimized. Another solution is to de-tune one of the interacting loops such that there is a difference in the integral (reset) time of 1 : 4. Hence one controller would act significantly quicker than the other, while the slower controller is acting more sluggishly.

If none of these options are possible, a decoupling control structure can be designed in theory, but an MPC control concept can also be used in practice.

Tutorial and Self-Study Questions

9.1 Explain the basic structure of cascade control.
9.2 Explain why cascade control is used.
9.3 Explain where and when cascade control is appropriate.
9.4 When would you recommend removing (or not implementing) cascade control?
9.5 Describe the tuning procedure for a cascade control system.
9.6 Draw the block diagram for cascade control.
9.7 Describe number of examples of cascade control.
9.8 Explain the basic structure of feedforward control.
9.9 Explain why feedforward control is used and where it is appropriate.
9.10 Draw the block diagram for feedforward control.
9.11 Explain how feedforward control is implemented in practice.
9.12 Describe a number of examples of feedforward control.
9.13 What process characteristics MUST be accounted for in feedforward control and how does this affect FF control action? What potential implications does this have for equipment design considerations?
9.14 Explain the basic structure of ratio control.
9.15 Explain how ratio control is implemented in practice.
9.16 Describe a number of examples of ratio control.
9.17 Congratulations! You have been recently hired by BYMTUI Engineering Ltd. as a junior process control engineer. Your supervisor recognizes that you should be well armed with new-found control knowledge, having read this book, and assigns you the task of implementing a control scheme for the flash distillation shown in the following schematic, Figure 9.11. Please draw using standard symbols your proposed control scheme on Figure 9.11.
9.18 The feed to the separator consists of a 50 : 50 mix of methanol and water and enters the separator at a flowrate of 100 kg-mol/h, at a pressure of 200 kPa, and at a temperature of 30 °C. Use HYSYS to simulate and test your proposed control scheme for this separator, which should achieve the objective of taking 40% of the feed overhead as a vapor product with the remaining 60% of the feed leaving the separator as a liquid product. Provide strip charts showing the overhead flowrate response with an overhead flowrate set-point change, and the overhead flowrate response for a set-point change of the feed flowrate.

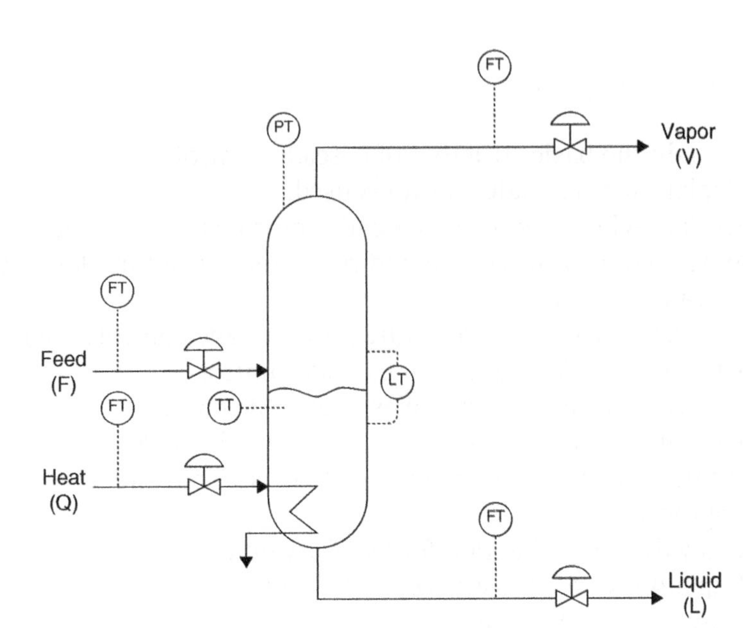

Figure 9.11 Flash distillation schematic.

References

Eckman, D.P. (1945). *Principles of Industrial Process Control*, Metered and Ratio Control, 194–199. New York: Wiley.

Murril, P.W. (1967). *Automatic Control of Processes*. Scranton, PA: International Textbook Company, (Cascade/Ratio, pp. 431–444, Feedforward Control, pp. 405–425).

Seborg, D.E. (1984). A perspective on advanced strategies for process control. *Modeling, Identification and Control* 15 (3): 179–189.

Shinskey, F.G. (1967). *Process Control Systems*. New York: McGraw-Hill, (Cascade/Ratio, pp. 154–60, Feedforward Control, pp. 204–229, Override Control, pp. 167–169).

Svrcek, W.Y., Mahoney, D.P., and Young, B.R. (2014). *A Real-Time Approach to Process Control*, 133–149. Wiley.

10 Model Predictive Control

10.1 Introduction to MPC

Multivariable model-predictive control (MPC) technology, generically referred to as "advanced process control" (APC), has been utilized by the process industries for over forty (40) years and has rightfully earned a place as high value-adding technology. The concept of MPC is an important process control concept that traces its initial commercial implementation back to petro-chemical refinery columns. From a theoretical (and practical) point of view, MPC accounts for the multivariable nature of most processes and how process disturbances propagate through the process over time. It also accounts for variations in process dynamics and is able to handle multiple simultaneous constraints. The great driving force for the development of MPC was that, in spite of the advances in control system capabilities, the base (advanced) regulatory control systems were unable to easily address the multivariable nature of most processes. APC offers solutions to these issues, as well as substantial economic benefits: savings of the order of 2–6% of annual operating costs typically (reported as high as up to 15%), and generates an extra 1% in revenue (e.g. Marlin et al. 1991; Brisk 2004).

In straightforward terms, MPC allows for coordinated efforts to control multiple and often interlinked aspects of a distillation process. Considering that a distillation column only has two real handles (mass and energy balance), an MPC often influence these handles to affect related parameters, such as product quality, product recovery rate, and throughput. The advantage of MPC over traditional advanced regulatory control (ARC) is MPC's ability to carry out process optimization, prior to

A Real-time Approach to Distillation Process Control, First edition. Brent R. Young, Michael A. Taube, and Isuru A. Udugama.

Companion website: www.wiley.com/go/Young/DistillationProcessControl

calculating the control moves that moves the process to its most optimal state (for the present conditions). A drawback of MPC is the relative complexity of setting up an MPC (to be discussed in detail in the following subsections) and the need for dedicated engineering support (in the form of an advanced process control expert). To this end, MPC should be only implemented when there is a clear economic business case. It should also be noted that MPC should be implemented on a solid foundation of both the regulatory and advanced regulatory controls.

10.2 To MPC or not to MPC

Even to this day, there is a debate about how APC and regulatory control, including ARC, should interact with each other. The debate centers on whether or not ARC should be bypassed when APC is in service and how much "regulatory control" functionality APC should contain, such as maintaining temperatures, levels, pressures and so on. To be sure, APC should never be considered as a replacement for fundamentally sound regulatory and ARC design. Regulatory control and ARC are the foundation upon which APC sits, therefore, in order for the "superstructure" to be sound, so must the foundation. Therefore, this section will describe guidelines for taking full advantage of modern control system capabilities with regulatory and ARC design and the design and scope of APC applications.

From a historical perspective, early APC developers had to make certain design choices regarding MVs and CVs due, in part, to the limited capabilities of control systems of the day, and the lack of proper regulatory control and ARC design, as described in previous sections.

As control system capabilities increased, however, APC designers failed to reassess their design philosophies to take advantage of the new control systems' capabilities and, regrettably, continued with the design choices of early implementations, eventually resulting in these design choices becoming "canon." This same design philosophy continues to be promulgated to the present day, even to the point of suggesting that regulatory controls and ARC be replaced with only MPC-based applications! The overarching supposition is that MPC will (or should) handle everything in terms of regulatory control and optimization, which is something for which APC was never intended. While APC has proven itself as a valuable profit-generating tool, it cannot be considered a "cure-all" or replacement for sound regulatory and ARC design. So, in terms of APC/ARC interaction, the following are submitted as guidelines for consideration:

1. APC CVs should focus on "hard" constraints, such as valve position, tower delta-pressure, and so on. Anything that represents a physical equipment limitation should be the only CVs in the scope of the APC application.
2. APC MVs should be limited to those "ultimate primary" regulatory and advanced regulatory control set points that an operator might otherwise manipulate, including quality targets.
3. Ensure that the underlying regulatory control and ARC structure help and support the APC application (rather than hinders or exacerbates it); if it does not, correct the underlying structure before implementing APC.
4. APC application execution times should be quite "slow" (relative to current practices), in the range of every 5–20 minutes or longer, to allow the underlying controls to address faster dynamic behaviors and higher frequency disturbances.

10.3 MPC Fundamentals

Model-predictive control (MPC) is a class of multivariable computer-based control schemes that use a process model to explicitly predict future plant behavior and calculate the appropriate control action required to drive the control variable(s) as close as possible to the optimal values. It is the most widely industrially used of all advanced control methods.

MPC can handle multivariable/interacting processes and challenging dynamics with relative ease by using a process model matrix that relates control variable responses to manipulated variables. It attempts to maintain measured and inferential variables either at a specific target or within a specified range. It also optimizes control effort to meet objectives and is potentially capable of handling all constraints (hard, soft, equality, and inequality constraints).

The general principles of MPC may be illustrated by considering first how the process variable will behave in the future if no further action is taken. The target control action to rectify what is left to be corrected after the full effects of the previously implemented control action. The four basic elements of MPC are as follows:

1. A reference or set-point trajectory specification, $y^*(k)$, where the argument (k) is represents the value of reference trajectory time series, y^* at the kth sampling or time instant.
2. Process variable prediction, $\hat{y}k + i$, where the argument $(k+i)$ represents the value of the predicted process variable, $\hat{y}$ at the $(k+i)$th sampling instant.
3. The model M predicts $\hat{y}k + i$, based on the manipulated variable or control action sequence, $u(k+j)$, where $u(k+j)$ is the value of the control action sequence at the $(k+i)$th time instant. The model, M, calculates the control action to satisfy an optimization objective, subject to pre-specified constraints. This is akin to using the model inverse, M^{-1}, which is mostly carried out numerically as the solution of an optimization problem.
4. The error prediction update,

$$e(k) = y_m(k) - \hat{y}(k) \tag{10.1}$$

Here $e(k)$ is the error prediction update at the kth sampling instant and $y_m(k)$ is the measured value of the process variable at the kth time instant.

These basic elements of MPC are illustrated in Figure 10.1.

Standard MPC uses linear models. That is, just like PID-based controls, MPC assumes that the process gain (KP) is constant across the

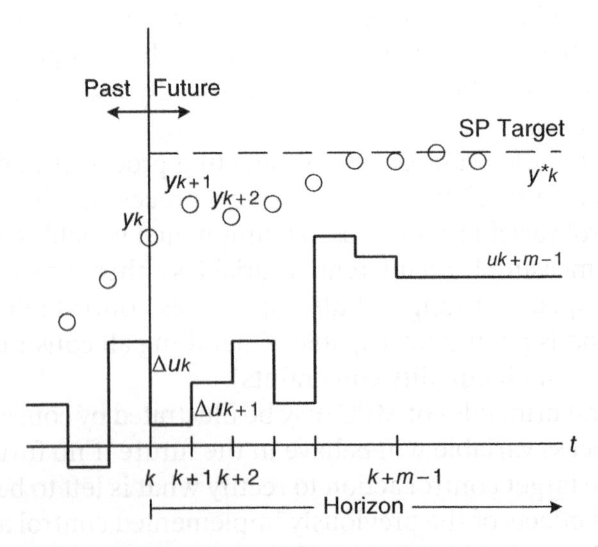

Figure 10.1 The basic elements of model predictive control.

entire operating range. Any nonlinear behavior is handled by transform functions, again, just as one would for regulatory controls. As MPC schemes are implemented on digital computers, linear discrete-time models are used. Three linear, discrete-time models are mostly used:

1. Finite convolution models
2. Discrete state space models
3. Transfer function models

Finite convolution discrete-time models are expressed in either impulse response or step response forms.

The impulse response model form shows the process variable sequence as a function of the control action sequence as a series of impulses like so:

$$y(k) = \Sigma g(i) \cdot u(k-i), \quad \text{for } i = 0 \text{ to } k \tag{10.2}$$

Here the parameters, $g(i)$, comprise the impulse response function.

The step response model form expresses the process variable output sequence as a function of the series of control action changes or steps as in the following equation:

$$y(k) = \Sigma b(\mathrm{i}) \cdot \Delta u(k-i), \quad \text{for } i = 0 \text{ to } k \tag{10.3}$$

Here $\Delta u(k) = u(k) - u(k-1)$ is the step or change in the control action between each sampling instant and the parameters, $b(i)$, constitute the step response function.

The impulse and step response functions are related as follows:

$$g(i) = b(i) - b(i-1) \tag{10.4}$$

$$b(i) = \Sigma g(j), \quad \text{for } j = 1 \text{ to } i \tag{10.5}$$

For real, causal systems because of a mandatory one-step delay:

$$g(0) = b(0) = 0 \tag{10.6}$$

The remainder of the response coefficients, $g(i)$ and $b(i)$, are obtained from noise-free data or other models. Identification of these coefficients is obtained via careful system identification of the plant and is a key requirement of MPC (as well as ARC/regulatory control).

Many factors contribute to the discrepancy between actual data and model predictions. These include unmodeled or unmeasured disturbances, fundamental errors in model structure, and/or unavoidable errors in model parameter estimates. The fundamental MPC strategy is pragmatic – it is assumed that the discrepancy is caused by unmeasured disturbances and is constant, so the current discrepancy to all predictions is an error prediction update.

Two industrial groups developed the first MPC techniques in the 1970s independently. Shell developed dynamic matrix control (DMC) in the United States (Cutler and Ramaker 1979), while a similar technique called model-predictive heuristic control (MPHC) was developed in France (Richalet et al. 1978). Both of these approaches were similar and shared many common features. Since then, many other related MPC products have been developed and installed worldwide. Qin et al. (1997) and Qin and and Badgwell (2000, 2003) and Morari and Lee (1999) have presented excellent reviews of MPC technology. DMC and its descendants remain the most popular of the MPC technologies and will be described hereon in Section 10.4.

10.4 Dynamic Matrix Control

This section will first introduce the concepts of DMC by considering the unconstrained case for a single process variable and then extend the treatment to include multiple variables and constraints. The chapter concludes with a brief consideration about some of the practical implementation aspects of MPC generally.

DMC uses a step response model to represent the process:

$$y'(t) = y(t_i) - y(t_0) = a_i \cdot \Delta u(t_0) \tag{10.7}$$

We will illustrate this model with a first-order plus dead time example where the process gain, $K_p = 1$, time constant, $\tau = 1$, and dead time, $L = 1$ (arbitrary units). A step change, $\Delta u(t)$, is introduced at time, $t = 0$, i.e. $\Delta u(0) = 1$. For a sampling time of $h = 1.0$, the step response coefficients can then easily be calculated, the first 5 of which are shown in Table 10.1.

Table 10.1 Example step response and step response model coefficients.

Time	*i*	$\Delta u(t)$	$y'(t)$	a_i
0.0	0	1	0	0
1.0	1	0	0	0
2.0	2	0	0.63	0.63
3.0	3	0	0.86	0.86
4.0	4	0	0.95	0.95
5.0	5	0	0.98	0.98

We can use the model of Eq. (10.7) to predict the process variable, $y(t)$, given the control moves, $\Delta u(t)$, as follows:

$$\begin{aligned} y(t_1)-y(t_0) &= a_1 \cdot \Delta u(t_0) \\ y(t_2)-y(t_0) &= a_2 \cdot \Delta u(t_0) + a_1 \cdot \Delta u(t_1) \\ \dots \quad & \quad \dots \\ y(t_n)-y(t_0) &= a_n \cdot \Delta u(t_0) + a_{n-1} \cdot \Delta u(t_1) + a_{n-2} \cdot \Delta u(t_2) + \cdots \end{aligned} \tag{10.8}$$

i.e. in vector notation:

$$y'(t) = y(t_i) - y(t_0) = \Delta a_i \cdot \Delta u(t_{n-i}), \quad \text{for } i = 1 \text{ to } n \tag{10.9}$$

In matrix form, we get the so-called dynamic matrix that DMC is named after:

$$\begin{vmatrix} y'(t_1) \\ y'(t_2) \\ y'(t_3) \\ \dots \\ y'(t_n) \end{vmatrix} = \begin{vmatrix} a_1 & 0 & 0 & \dots & 0 \\ a_2 & a_2 & 0 & \dots & 0 \\ a_3 & a_2 & a_1 & \dots & 0 \\ & \dots & & & \\ a_m & a_m & a_m & \dots & a_1 \end{vmatrix} \begin{vmatrix} \Delta u(t_0) \\ \Delta u(t_1) \\ \Delta u(t_2) \\ \dots \\ \Delta u(t_{n-1}) \end{vmatrix} \tag{10.10}$$

$$\text{i.e. } y' = A \cdot \Delta u \tag{10.11}$$

Here we are assuming that $n > m$, where n is the prediction horizon and m is the model horizon. This is often the case.

Using the dynamic matrix, we can predict the process variable, y, for a series of manipulated variable moves, Δu. This is then implemented in DMC as a moving horizon algorithm. All past y and u values are known, and future u values are chosen to regulate y using the step response model and past u values. This is then repeated after another time interval, i.e. only first Δu is actually implemented.

The prediction vector, y^p ($y(t)$ for $t > t_0$), is the predicted value of the process variable assuming no future manipulated moves, i.e. $\Delta u(t) = 0$, $t \geq t_0$, and it is obtained from the following calculation:

$$\begin{vmatrix} y^p(t_1) \\ y^p(t_2) \\ \cdots \\ y^p(t_n) \end{vmatrix} = \begin{vmatrix} y(t_{-m}) \\ y(t_{-m}) \\ \cdots \\ y(t_{-m}) \end{vmatrix} + \begin{vmatrix} a_m & a_m & a_{m-1} & \cdots & a_3 & a_2 \\ a_m & a_m & a_{m-1} & \cdots & a_4 & a_3 \\ & & & \vdots & & \\ a_m & a_m & a_m & \cdots & a_m & a_m \end{vmatrix} \begin{vmatrix} \Delta u(t_{-m}) \\ \Delta u(t_{-m+1}) \\ \cdots \\ \Delta u(t_{-1}) \end{vmatrix} \tag{10.12}$$

$$\text{i.e.} \quad \underline{y^p} = y(t_{-m}) + A^p \cdot \Delta u^p \tag{10.13}$$

When combining with future control moves, we get the following equation:

$$y = y^p + A \cdot \Delta u \tag{10.14}$$

The prediction of the process variable from Eq. (10.14) will result in errors due to the following:

1. Errors in step response model coefficients
2. Unmeasured disturbances
3. Nonlinear behavior
4. Not being at steady state at $t = t_{-m}$. (or not knowing the "recent history" of the process)

DMC uses a simple prediction error to account for these errors, as follows:

$$e = y(t_0) - y^P(t_0) \tag{10.15}$$

This results in a revised prediction of the process variable, as follows:

$$y = y^p + A \cdot \Delta u + e^T \tag{10.16}$$

Here $e = [e\ e\ \ldots\ e]$.

DMC is based on minimizing the error from the set-point trajectory. The objective function is the sum of the square of the errors from the set-point trajectory n steps into the future:

$$\Phi = \Sigma\left[y_{sp} - y^P\left(t_i\right)\right]^2, \quad \text{for } i = 1 \text{ to } n \tag{10.17}$$

Here y_{sp} is the set-point trajectory and n is the prediction horizon.

We then define:

$$E\left(t_i\right) = y_{sp} - y^P\left(t_i\right) - e \tag{10.18}$$

If Eq. (10.18) is substituted into Eq. (10.17), we arrive at the following revised form of the objective function:

$$\Phi = \Sigma\left[E\left(t_i\right) - y_c\left(t_i\right)\right]^2, \quad \text{for } i = 1 \text{ to } n \tag{10.19}$$

Here

$$y_c = A \cdot \Delta u \tag{10.20}$$

The objective of the DMC controller is to choose $\Delta u(t_i)$, for n moves into the future to minimize Φ, the sum of the squares of the errors from set point. The control law is obtained by differentiating Φ with respect to Δu:

$$\frac{d\Phi}{d\Delta u} = A^T\left(A \cdot \Delta u - \Phi\right) = 0 \tag{10.21}$$

Solving Δu, we obtain:

$$\Delta u = \left(A^T \cdot A\right)^{-1} \cdot A^T \cdot E \tag{10.22}$$

If A is square, $(A^T \cdot A)^{-1} \cdot A^T = A^{-1}$.

However, we are not quite yet done with the control law. It turns out that Eq. (10.22) results an overly aggressive response and $(A^T \cdot A)^{-1}$ can well be mathematically ill-conditioned due to process model mismatch or incorrect identification of process dead time. As a result, a diagonal move suppression matrix is employed to overcome these issues:

$$Q = qI \tag{10.23}$$

Here the move suppression factor, q, is positive. A large value penalizes Φ more for control moves and is also used for large nonlinearities/disturbances.

The resulting control law is:

$$\Delta u = \left(A^T \cdot A + Q^2\right)^{-1} \cdot A^T \cdot E \tag{10.24}$$

From the above discussion, the reader will have noted that the DMC controller has a large number of tuning parameters – the model horizon (m), the prediction horizon (n), the number of control actions, the model parameters (a_i), and the move suppression matrix (Q). The prediction horizon (n) must be chosen, so the results of control actions can be observed within it. The number of control actions calculated is dictated by process dead time and the prediction horizon.

Once these tuning parameters are defined, DMC is implemented as follows. The controller gain matrix, K_c, is first calculated off-line via:

$$K_c = \left(A^T \cdot A + Q^2\right)^{-1} A^T \tag{10.25}$$

Then the uncontrolled output predictions are calculated online via:

$$y^P = y\left(t_{-m}\right) + A^P \cdot \Delta u^P \tag{10.26}$$

And then finally the next control action (only) from the control law is calculated online via:

$$\Delta u = \Sigma k_{ii}\left[y_{sp}\left(t_i\right) - y^P\left(t_i\right)\right] \tag{10.27}$$

Here the k_{ii} are the elements of the first row of K_c.

Thus far we have developed the case for DMC of a single-process variable without constraints. However, as mentioned by several authors, the value of MPC is its ability to handle multiple interacting variables and multiple constraints. Therefore, we will consider extension to multiple variables and constraints next.

Extension of DMC to multiple input, multiple output (MIMO) processes is handled by using augmented vectors and matrices. For example, the augmented matrix and control moves vector are as follows:

$$A = \begin{vmatrix} A_{11}A_{12}\cdots A_{1j} \\ A_{21}A_{22}\cdots A_{2j} \\ \cdots \\ A_{k1}A_{k2}\cdots A_{kj} \end{vmatrix} \& \ \Delta u = \begin{vmatrix} \Delta u_1 \\ \Delta u_2 \\ \cdots \\ \Delta u_j \end{vmatrix} \tag{10.28}$$

Here j is the number of manipulated variables and k is the number of control variables.

Consequently, we also need to prioritize the control objectives in the control law. This is done by adding an augmented diagonal weighting matrix, W, that contains diagonal matrices, which provide the relative weighting for the controlled variables:

$$\Delta u = \left(A^T \cdot W^2 \cdot A + Q^2\right)^{-1} A^T \cdot W^2 \cdot E \tag{10.29}$$

Therefore, the augmented diagonal weighting matrix, $W = w_i \cdot I$, becomes an additional set of tuning parameters for the multivariable controller (i.e. the relative weighting factors for the controlled variables).

So far, we have assumed that the control moves, Δu, are unbounded. However, this is rarely the case in real processes. The first approach to solving a constrained MPC case was a modified unconstrained approach to DMC, which is as follows:

1. Solve the unconstrained DMC problem,
2. Check for constraint violations,
3. Fix any infeasible variables from the above solution at their limits, remove them from the DMC problem, and resolve,
4. Repeat steps 2 and 3 until a feasible solution found

The problems with this modified unconstrained DMC approach are that, whenever constraints are violated, the dynamic matrix is modified. The entire DMC problem must be solved online. It is an iterative approach, and a solution is not guaranteed.

The solution to these problems was the development of the Quadratic DMC approach (QDMC) (Qin and and Badgwell 2000, 2003). This is an optimization-based approach with the following objective function:

$$\min \Phi = \left[y_{sp} - y^P(t_i)\right] W^2 \left[y_{sp} - y^P(t_i)\right] + \Delta u^T Q^2 \Delta u \quad (10.30)$$

This is objective function is subject to the following constraints:

$$y = y^P + \underline{A\delta u} \quad (\text{model}) \quad (10.31)$$

$$g(y, \Delta u, u) > 0 \quad (\text{constraints}) \quad (10.32)$$

$$u_L < u < u_H \quad (\text{control move bounds}) \quad (10.33)$$

$$\Delta u_L < \Delta u < \Delta u_H \quad (\text{control move rate bounds}) \quad (10.34)$$

$$y_L < y < y_H \quad (\text{process variable bounds}) \quad (10.35)$$

The major advantages of QDMC are that any constraint can be explicitly represented in the control calculations, and it determines the optimal, feasible moves in the prediction horizon. However, there is no analytical solution – the optimization must be solved at each control interval. A Quadratic Program does this, hence the name Quadratic DMC.

Thus far we have largely, by implication, considered the number of inputs and outputs in the DMC formulation to be the same (i.e. the case of square DMC). However, the case when the number of inputs and outputs are not equal occurs frequently in real processes. This is termed non-square DMC. In the case of non-square DMC, steady-state "economic" optimization is used to specify good compromises among variables. For excess inputs, the optimization is often specified as a linear program. The solution to this problem provides steady-state targets to which input variables are driven. Steady-state targets are then introduced into QDMC; the problem is then solved online and is computationally heavy.

10.5 Setting Up a MPC in Distillation

In this subsection, an example case will be presented and the steps, described above, followed in implementing a predictive model controller on an industrial high-purity multicomponent distillation column. The example used in this set-up is the same column as introduced in

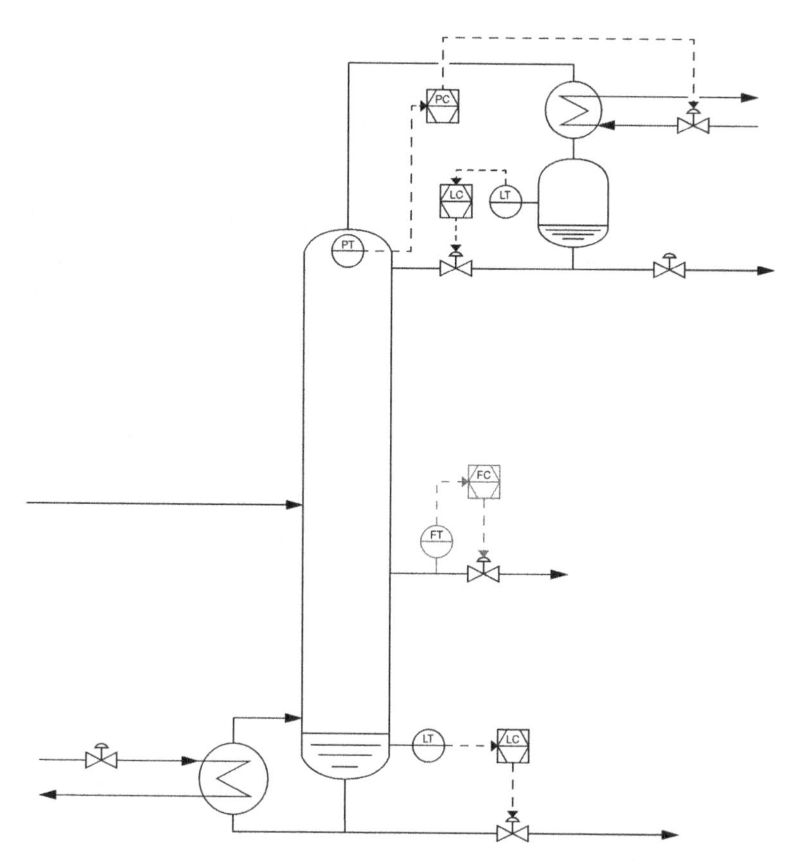

Figure 10.2 Side-draw column example.

Section 8.5, which is an industrial methanol tower with a feed containing methanol, water, and ppm levels of ethanol. This stream is refined in a distillation column that has a small side draw to achieve high-purity product and bottoms (ppm levels), as well as high recovery rates. Figure 10.2 illustrates the base control structure that is implemented on this particular distillation column.

Prior to moving forward with implementing a MPC, the need for, and the expected benefits of, implementing a MPC solution must be established.

In this particular case study, the following argument is made for the implementation of a MPC:

- The product specification must be maintained at <10 ppm of ethanol. As described previously, this will inherently exhibit a nonlinear behavior due to the ultrahigh product purity specification;

- The product recovery must be >98% to ensure that no methanol is present in the bottoms' stream;
- The reboiler steam is used sparingly – only enough to maintain the first 2 objectives.

However, prior to moving further, it is important to understand that further modifications are required to both the underlying control structure and measurements, which will allow an MPC to function efficiently in this particular situation.

- The product purity must be actively monitored *in situ*. In this instance, this requires the installation of an online gas chromatograph.
- A closed-loop (regulatory) control structure is required to manipulate the distillate ethanol flow rate. This requires the installation of a flow element and transmitter.
- A calculation of the product recovery rate is required. In theory, this can be carried out by calculating the ratio of mass flow rate of methanol in the feed and the distillate flow rate (which is >99.9% methanol). More cost-efficient practical solutions can be determined as detailed in Cutler and Ramaker (1979).
- A closed-loop (regulatory) control structure is required to manipulate the reboiler duty stream. This requires the installation of a flow element and transmitter.

To this end, the following base regulatory control structure and measurements must be implemented and validated prior to considering MPC.

10.5.1 Model Setup

Based on the above structure, the following 2×2 matrix (and the subsequent process model requirements) are given in Table 10.2.

Table 10.2 MPC process model requirements for MPC example.

	MV_1 (distillate flow rate)	MV_2 (reboiler steam flow rate)
PV_1 (distillate ethanol concentration)	G_{11}	G_{12}
PV_2 (product recovery rate)	G_{21}	G_{22}

The transfer function-based process models can be determined through step tests. Depending on the type of MPC, the exact method to input these step test may vary. For example, some commercial MPC applications may only allow a simple first-order plus time delay (FOPTD) to be implemented for each model transfer function.

Once the process models have been finalized, model-validity ranges and the desired set point or operating range of each control variable must be established. In this case they are:

- Product recovery rate range and set point
- Distillate ethanol concentration range and set point

Similarly, the range of the manipulated variables is also needed to ensure the subsequent MPC tuning steps are carried out appropriately. In this case, they are:

- Reboiler steam flow rate range
- Distillate flow rate range

10.5.2 Objective Function

The simplest objective function that can be set for an MPC is to ensure the set points of the process variables are met with the minimum movement of the manipulated variables as detailed in Eq. (10.30). The relative weight for the two control variables in this case can be adjusted by changing their ranges. However, in more advanced formulation of an MPC, an explicit economic cost function can be written as the overall objective function. This objective function will account for the set points and other aspects of economic importance. For example, in this particular case study, such an objective function simplistically may look like:

$$\text{Min } \Phi = M \times \text{product recovery rate} - N \times \text{reboler duty} \tag{10.36}$$

where M will be the economic value (per ton) of recovering more product, while N will be the cost of reboiler duty. In this type of objective function formulation, strict constraints will also need to be introduced. And, in this situation, the product ethanol concentration of <10 ppm should be set up as a hard constraint that must be guaranteed. In either situation, control variable set points are required.

10.5.3 Tuning

As discussed previously in this chapter, the tuning of an MPC application involves multiple parameters. Like a PID controller, appropriate tuning on an MPC can be seen more like "art" rather than a rule-based approach. Moreover, each MPC application/product may have different parameters that must be tuned. Hence, rather than describing steadfast rules, this book will discuss each common parameter and its effect on the control actions that the MPC will take.

The first parameter that must be set in an MPC application is the k value, which is the execution interval. Having a smaller k value means the MPC controller is executed more frequently. From a theoretical point of view, the "Shannon's Sampling Theorem" suggests the sampling (hence the execution) interval between 1/5 and 1/10 of the shortest time constant. In general, the authors suggest a $k > 20$ minutes. This follows based on the premise that the MPC is intended to address "slow" aspects of the process' behavior and that the underlying ARC handles the "fast" behavior.

The next variable that must be chosen is the predication horizon, P, which is usually the number of steps into the future the process behavior is predicted. The authors suggest that the number of prediction steps be set such that it is, at least, 3–4 times larger than the most dominate process time constant. Combining this information with the k value, the minimum the p value should be set is 5 (in case there is only one process time constant of importance and the sampling rate is set at 1/5 the time constant). The smaller the p value, the more accurate the prediction will be.

The control horizon, m, determines the number of control periods (moves) that the MPC will optimize when predicting the output of the MPC. A value of for m means the MPC optimization algorithm will only consider that a "single" move can be made for the prediction period. Hence, a lower m means the optimization does not consider that future coordinated control moves will be made, rather trying to "remedy" any off-specification variables in a single move. The m value should be somewhat smaller than the p value. By default, an m value of 2 can be chosen as this means the optimization algorithm in an MPC considers the next move that can be made when predicting the current control move. This should in principle lead to a more "measured" set of control moves.

The last set of parameters that must also be set is the weight matrices that determine the relative importance between variations in control variable Y and the manipulated variable U, collectively referred to as the ΓU and ΓY matrices. The weight of ΓU and ΓY must add up to 1. By default, ΓU and ΓY can be kept at 0.5, meaning equal weight is given to having stable Y and U values. However, if significant swings in the manipulated variable can be tolerated, then ΓY can be set close to 1 (not at 1), while ΓU by default will be close to zero. This type of setting means the MPC would react as aggressively as possible, without consideration for oscillations in the manipulated variables, U, to ensure aggressive set-point tracking and no variations are affecting the control variables. Similarly, a ΓU close to 1 means the MPC would put a higher weight on ensuring smooth manipulated variable actions when no process variations affect the control variables. This type of setting would most likely be implemented if aggressive movements in manipulated variables may have other ramifications in the plant. For example, if reboiler steam is a manipulated variable, then "aggressively" manipulating this variable can have ramifications plant-wide due to the fact that steam is a utility that is used across the facility and any "aggressive" movement by one or a few major users could adversely affect the entire steam supply.

10.6 Digitalization and MPC

MPC technologies can benefit from the increased emphasis on digitalization developments both in industry and academia. On the one hand, with the emergence of digital twins (accurate validated digital representation of a physical process), there is the ability to improve the prediction models used in an MPC application. More specifically, it is possible to improve upon the classic transfer function models currently used and replace them with improved first principle-based models derived from digital twins. On the other hand, with improved data-driven approaches being available, there is also the possibility of developing hybrid process modeling approaches that can update the transfer function models used in MPC while in operations on a live plant. It is also important to note that a long body of research work is already available in academia for both of these trends. However, as occurs with much academic research, these developments are still not implemented in mass across industry.

Tutorial and Self-Study Questions

10.1 What are the pre-requisites for implementing MPC control and when should MPC controls be considered?

10.2 What are the pre-requisites for implementing MPC control and when should MPC controls be considered?

10.3 What is the "purpose in life" of EVERY MPC/APC application? Given this, what is the fastest execution rate that an MPC/APC application should nominally run?

10.4 What is the relationship between model-predictive control (MPC/APC) and (advanced) regulatory controls? Given this relationship, what should an MPC/APC application NEVER be forced to do?

References

Brisk, M.L. (2004). Process control: potential benefits and wasted opportunities. *Proceedings, 5th Asian Control Conference* (20–23 July 2004), Melbourne, Australia, Vol. 1, pp. 10–16.

Cutler, C.R. and Ramaker, B.L. (1979). Dynamic matrix control – a computer control algorithm. *AIChE 86th Spring National Meeting*, Houston, TX (April 1979).

Marlin, T.E., Perkins, J.D., Barton, G.W., and Brisk, M.L. (1991). Benefits from process control: results from a Joint Industry-University Study. *Journal of Process Control* 1 (2): 68–83.

Morari, M. and Lee, J.H. (1999). Model predictive control: past present and future. *Computers and Chemical Engineering* 23: 667–682.

Qin, S.J. and Badgwell, T.A. (1996). An overview of industrial model predictive control technology. *Proceedings, Chemical Process Control V*, Tahoe City, CA (1996), *AIChE Symposium Series* No. 316, Vol. 93, 1997, CAChE and AIChE, New York, pp. 232–256.

Qin, S.J. and Badgwell, T.A. (2000). An overview of nonlinear model predictive control applications. In: *Nonlinear Model Predictive Control*, Progress in Systems and Control Theory, vol. 26 (ed. J.C. Kantor, C.E. Garcia and B. Carnahan), 369–392. Basle, Switzerland: Birkhauser Verlag.

Qin, S.J. and Badgwell, T.A. (2003). An overview of industrial model predictive control technology. *Control Engineering Practice* 11 (7): 733–764.

Richalet, J., Rault, A., Testud, J.L., and Papon, J. (1978). Model predictive heuristic control: applications to industrial processes. *Automatica* 14 (5): 413–428.

11 Plant-Wide Control in Distillation

Distillation is often part the final set of unit operations of a production process tasked with ensuring on specification production of a finished product. So far in this book, we have focused on controlling distillation unit operations that typically consisted of a single distillation column and, in the latter chapters, distillation configurations that had two integrated columns. However, taking a broader perspective, it will be seen that there is a need to widen the focus of one's control efforts to identify if control actions at multiple unit operations need to be coordinated so as to ensure optimal and robust overall operations.

It should be noted that the design of an overall production process focuses on the economic bottom line (i.e. developing design and integration solutions) that results in:

- Lower capital costs by reducing the equipment required to carry out product synthesis and refinement;
- Lower energy and operating costs due to better utilization of heat and materials flows (e.g. baffled reactors);
- Higher production rates due to the continuous *in situ* removal of inert or inhibitory material, when reaction and separation are combined into one unit operation.

These aspects further increase the need for coordinated control efforts across unit operations. In practice, these types of plant-wide considerations often warrant a detailed look at a given situation on a case-by-case basis. This is because both the underlying process design

A Real-time Approach to Distillation Process Control, First edition. Brent R. Young, Michael A. Taube, and Isuru A. Udugama.

Companion website: www.wiley.com/go/Young/DistillationProcessControl

and the economic aspects of the overall plant operations create a unique set of needs, wants, and limitations that the control structures must accommodate. The focus of this chapter is not to look at these nuances, but rather to focus on the burden plant-wide considerations place on the controls design at the distillation column level. As such, this chapter will focus on empowering the reader to "think" and identify the potential issues and, where practical, point out some generic mitigating strategies in terms of controls design.

11.1 Distillation Column Trains

Even within distillation, it is common to have multiple distillation columns operated in series: both in fine chemicals and refining process operations, as shown in Figure 11.1 (e.g., Buckley 1964; Shinskey 1996). The reason for such an arrangement can include all of the above-mentioned economic considerations, but there may also be thermodynamic reasons, such as the need for multicomponent separations and/or high purity separations. To minimize the design and construction costs, there are no buffer tanks built between these columns. As such, any variations in the front of the distillation train will affect the operations of the downstream columns (forward coupled).

From a control perspective, this type of forward coupling means that accumulation of variations through each column makes the on-specification control of the columns downstream all the more challenging. To address this issue, the following concepts are suggested.

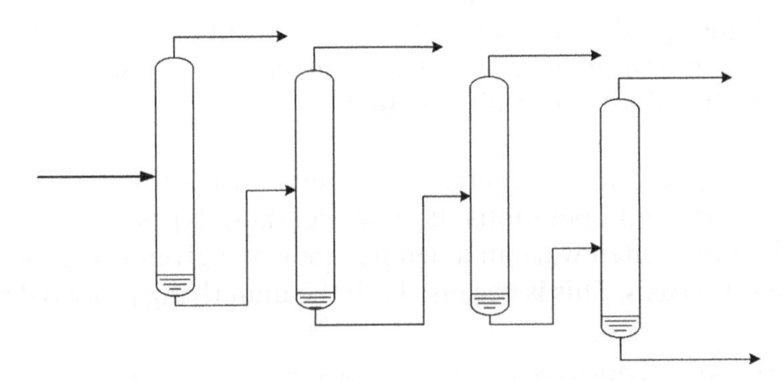

Figure 11.1 Schematic of multiple distillation columns operated in series.

11.1.1 Average Flow Control

One way of reducing the variations impacting the downstream columns that has been suggested in the literature is to use average-level control (e.g. Shinskey 1997; Svrcek et al. 2014) on the bottoms flow, as shown in Figure 11.2. This type of control structure ensures steady and smooth feed flow to the next distillation column, thus maintaining a consistent flow throughout the distillation train. However, a consequence of averaging level control is that the bottoms accumulator level is no longer strictly controlled, as the bottoms flow is typically used to maintain the bottoms liquid inventory. Instead, the bottoms accumulator level will vary as the set point to the flow controller is updated only to ensure that the bottoms accumulator does not become empty or overflow (e.g. it is allowed to fluctuate between 75% and 25% with limited changes to the flow set point).

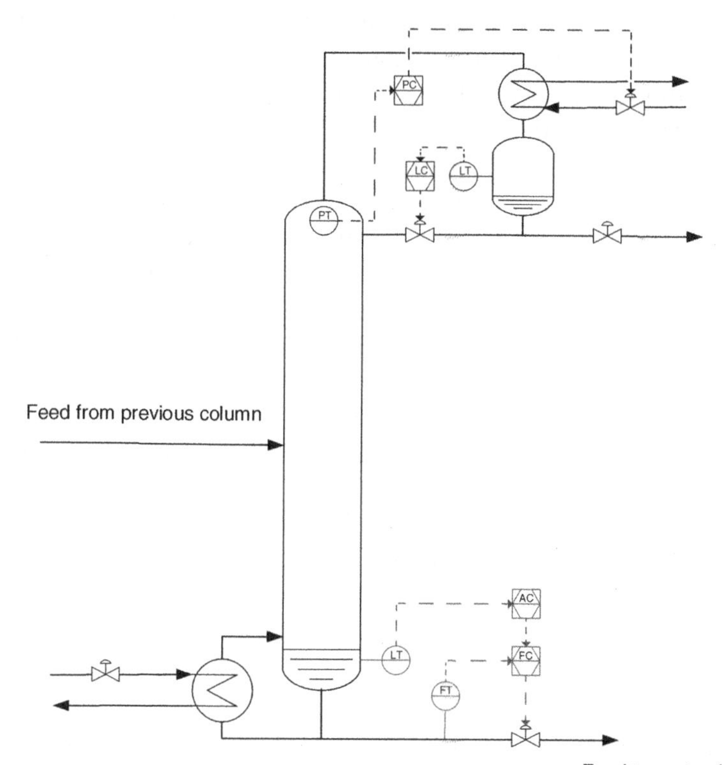

Figure 11.2 Averaging flow control on the bottoms flow to reduce the variations.

A consequence of this control scheme is that the bottoms accumulator is now treated as a "buffer tank" between each of the columns by using the liquid hold-up capacity of the bottoms sump in each distillation column. But, as discussed in Chapter 4, distillation tower accumulators are not designed with significant surge capacity and implementing this control scheme has the undesirable side effect of changing the perceived behavior of the tower due to artificially altering the steady state and dynamic responses to MV changes. Thus, it is recommended to explore other potential solutions before implementing this approach.

11.1.2 Alternatives to Average-Level Control

The alternatives to average-level control are as follows:

- A feed-forward strategy may be implemented, as described in Chapter 9, where the reboiler duty is adjusted such that the energy balance in the column is maintained, thus ensuring on-specification production. Similarly, the distillate flow rate may also be manipulated through a feed-forward control mechanism to maintain the tower's mass balance.
- The use of one distillation column in the overall distillation column train as a disturbance dump: In this case, a single distillation column in the train may be designed to ensure smooth and on-specification feed flow is delivered to the downstream columns. The control objective for the columns designated as such focuses on absorbing disturbances, rather than the production on on-specification distillate. As such, the distillate product of this column would be allowed to vary to "absorb" disturbances.
- Model-predictive control where MVs and CVs are looking across multiple distillation column unit operations may also be used to take "coordinated" control actions such that disturbances are buffered. However, this also requires the development of a large and potentially complex process model of the affected columns, if not the whole distillation train.

11.2 Heat Integration (Energy Recycle)

Heat integration is another common practice in petrochemical processes. Primarily, heat integration is carried out to reduce the cost of heating (and cooling) of processes. With distillation being a key energy consumer, heat

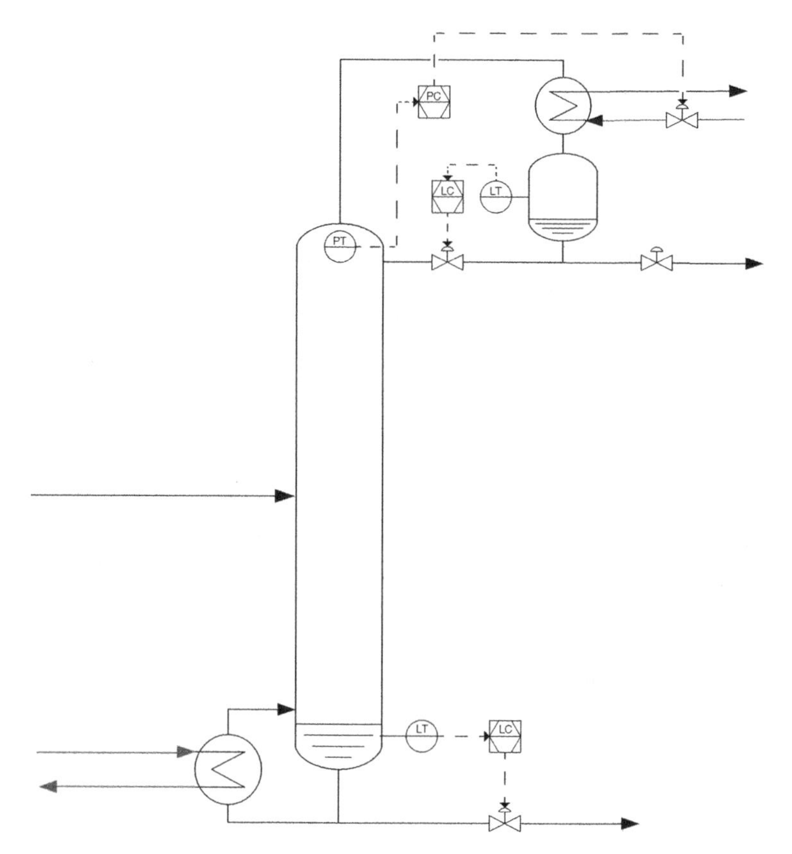

Figure 11.3 Waste heat reboiler.

integration is often practiced in various forms. The simplest example of heat integration is shown in Figure 11.3, where the reboiler duty is generated by waste heat from another production process (such as reactor). At first glance, one may assume that there is no influence by the upstream unit operations on this column's operations and control. However, a closer inspection reveals that the steam flow (reboiler duty) is no longer a manipulate-able variable. Consequently, it is no longer possible to independently manipulate the steam flow rate as steam production is now dependent on another unit operation. In fact, reboiler duty (steam flow) can now be classified as a disturbance variable since the steam is produced through waste heat recovery. Thus, as with all heat integration endeavors,

- Any advanced regulatory controls developed can no longer use reboiler duty as a manipulated variable, resulting in a loss of degrees of freedom.

- Reboiler duty itself will now propagate variations in downstream unit operations.

11.2.1 Auxiliary Steam Boilers

In practice, the example shown in Figure 11.3 often requires the installation of auxiliary steam boilers that are used for a "load balancing" system, as well as for startup conditions (where the auxiliary heat source may not be available), to ensure that during normal operations the steam flow rate to the distillation column can be manipulated within an acceptable range.

11.2.2 Feed Preheating

Another common type of heat integration in distillation columns that also requires a "re-think" is feed preheating. There are two (2) types of general styles of preheater designs:

1. One that uses a process stream from another unit (such as a pump around, as described in Chapter 6), or
2. One that uses the bottoms stream of the subject distillation tower.

While both options are a very common design practice, there are some caveats regarding mechanical and controls designs and the resulting operational effects they have on the tower.

There are two common designs for feed preheaters: those that employ a bypass on the "other" stream and those that do not, as illustrated in Figure 11.4a, b, respectively. While a steady-state material and energy balance presuppose a static set of process conditions for all process streams, the operational reality in a running plant is anything but static!

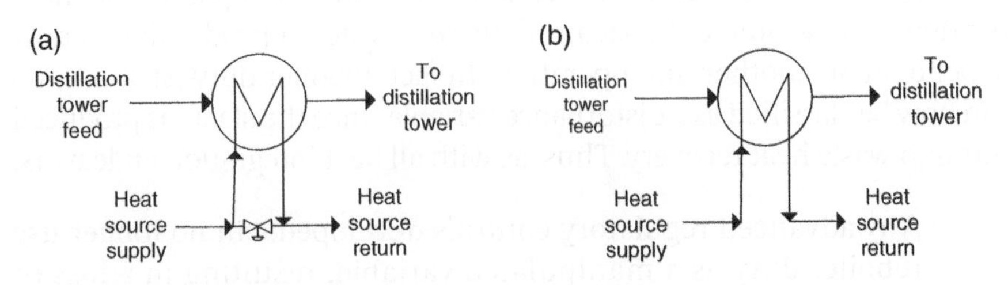

Figure 11.4 Distillation tower feed preheater designs.

This fact, however, is sometimes lost with plant designers and often results with the design shown in Figure 11.4b. To understand the effect on the operations and, thus, the control of the tower, another thought experiment is warranted.

Before getting into the thought experiment, one must identify the potential disturbance sources associated with the addition of the preheater and their effects on the tower feed stream, as well as what this means as far as new or additional control objectives are concerned. While the disturbance sources may be somewhat apparent, they will be enumerated and their effects discussed in some detail. However, to give these effects context, one must first explore the impact of a feed preheater on the tower's functioning and, thus, the possible controls that one may want to implement to affect that functioning.

When designing a new or assessing an existing distillation tower, some of the many parameters that must be in hand are the feed flow rate, its composition, and its enthalpy; that is the incoming mass and energy to the column. If any of these three parameters change, then the operations of the tower changes with them. To be clear, all three of these parameters are affected by equipment and controls upstream of this column, and thus outside the scope of this tower's controls. However, mitigating their effect(s) is very much within the scope of this tower's controls. Since the use of a feed preheater has been included in the design, it will be very prudent to assess how this additional equipment can be used to help ensure the column achieves its design and operational goals.

Furthermore, while changes to these tower feed properties may be significant, thus requiring this tower's controls to use them as feed forward variables (assuming that reliable measurements exist), for the present discussion, they will be relegated to the domain of feedback control. With all the above in mind, it is now possible to explore the disturbance sources resulting from the addition of the preheater and their potential effects to the tower feed stream's enthalpy. The disturbance sources resulting from the addition of the preheater are as follows:

1. Heat source (the "other stream") total flow rate.
2. Heat source temperature.
3. A somewhat less significant, but nonetheless potential, disturbance is the heat transfer capacity of the exchanger. This manifests in two ways:
   i. Fouling or plugging (mechanical)
   ii. Laminar flow conditions (low flow velocity)

For further clarity, the above items will be referred to as the exchanger's "input variables" to distinguish them from corresponding parameters of the tower feed stream. To begin the thought experiment, let's assume a steady-state condition where all thermal and flow disturbances have passed (as one is given by a steady-state simulation). If the mechanical design of the preheater is as shown in Figure 11.4b, then a change to any of the exchanger's input variables will affect the tower feed stream's enthalpy, thus affecting the behavior of the fluid as it enters the tower. In terms of impact from changes to the exchanger's input variables, the relative magnitude of each variable is as follows:

1. Heat Source Flow Rate

   In general, this has the largest impact on the tower feed enthalpy. It is also under the direct control of another part of the process and subject to changing based on objectives that are separate and independent from this tower's objectives. Thus, this measurement represents the most critical variable to monitor and, if possible, to provide a means of affecting the fluid flow that is fed into the exchanger.

2. Heat Source Temperature

   While this parameter provides the driving force for heat transfer, its impact is potentially smaller than, say, the heat source flow rate, but still significant. In particular, it is the temperature difference (ΔT) between the heat source and the tower feed stream that is important. Therefore, monitoring the ΔT is critical, especially if either or both are subject to diurnal or other large and/or frequent disturbances.

3. Heat Transfer/Flow Velocity

   This variable represents a few challenges in as much as it is not measured directly, but rather, is inferred or calculated, based on other measurements. Nevertheless, it represents a potentially critical variable, depending on the fluid's physical properties: if the fluid is sourced from, say, a crude vacuum tower bottoms, then maintaining a minimum velocity and temperature (within and after the exchanger) is vital to ensure that the fluid does not solidify inside the exchanger.

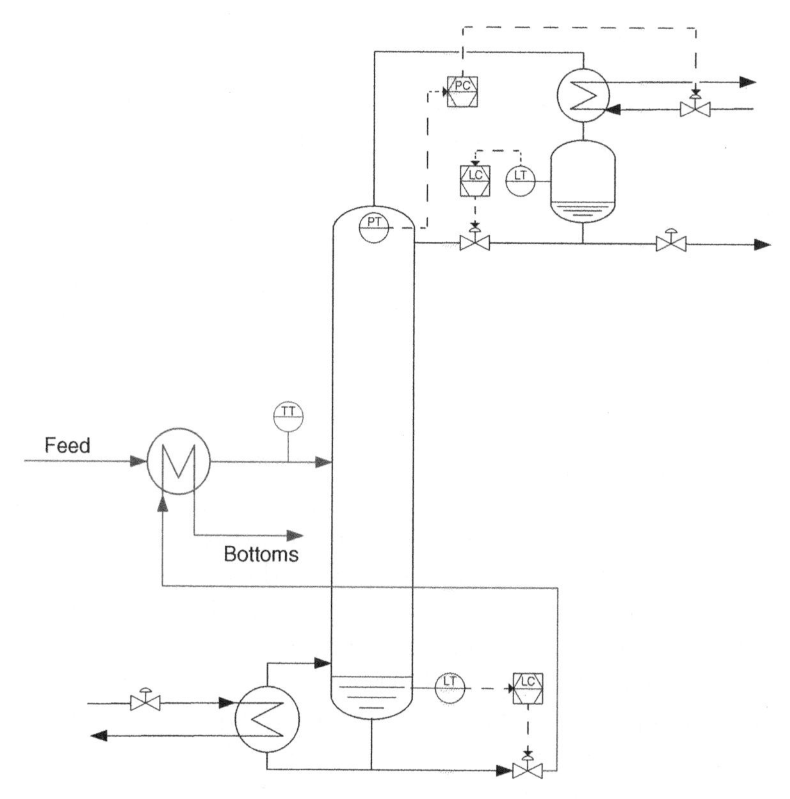

Figure 11.5 Feed preheating.

At the beginning of this section, it was indicated that the "other stream" could be the bottoms stream of the distillation tower for which preheat is provided, as illustrated in Figure 11.5. Before concluding this discussion, however, this variation must be examined closer with its own thought experiment.

As before, starting at some steady-state condition, one must consider how the exchanger's input variables are affected by the tower's operations. Based on observations made above, it should be apparent that a change to the reboiler duty will affect the bottom's stream temperature, which is one of the feed preheater's input variables. As this change progresses through the preheater, it results in affecting the tower feed stream's enthalpy. This, in turn, affects how the incoming fluid behaves upon entering the tower, thus causing a change in the vapor–liquid traffic, which affects the tower's energy balance, potentially then requiring a change to the reboiler duty, reflux flow, and/or tower mass balance in response: either increasing or decreasing, all subject to the VLE characteristics of the feed components and the tower's controls configuration. The net effect is that of a self-reinforcing "positive feedback loop" (e.g. Luyben 1993a,b,c; Luyben and co-workers 1997, 1998; Svrcek et al. 2014; Tyreus and Luyben 1993).

Given these facts and observations, along with the realization that nothing in a running plant is or remains static, it should now be obvious that a robust design for a distillation tower feed preheater should look like that shown in Figure 11.4a. To put a finer point to this, the following rules are submitted regarding the mechanical and controls design for feed preheaters:

1. First and foremost, avoid being overly obsessed with getting "everything available" from the "other stream": Both the tower feed and the "other stream," and their respective process conditions, are subject to change. Therefore, the prudent engineer will ensure that the means to respond to these changes are included in the mechanical and controls design.
2. Provide a means to bypass the preheater; either the tower feed or the "other stream," whichever stream is bypassed, ensure that a flow indication for the actual fluid going through the preheater is included. This ensures that any minimum flow velocity and/or heat transfer requirements can be accurately monitored.
3. Provide inlet and outlet temperature indications for both shell-side and tubeside streams; if any signal is (or may be) used for control, avoid thermocouples and utilize RTD/transmitter devices (to ensure appropriate signal fidelity and accuracy).
4. While the exchanger may be designed to capture 100% of the available heat, design the controls to capture no more than 70–80% of available energy to provide control flexibility and capacity to make dynamic adjustments.

In order to provide further clarity on the control aspects associated with feed preheaters, a few detailed design examples are presented here. These examples are presented in increasing order of robustness and, thus, cost; from simplest to "complex," but illustrate the thought process every process design, as well as process control, engineer should employ when assessing such equipment designs. While other types of heat integration are possible, such as using one tower's overhead vapor as preheat and condenser or reboiler heat source, but require far greater analysis and is outside the scope of this document. The "basic" preheater controls designs are as follows:

1. Simple feed temperature control
2. Tower feed supply temperature feed forward
3. "Other stream" supply temperature feed forward
4. "Other stream" exchanger feed flow or "duty" control

11.2.3 High-pressure/Low-pressure Columns

The use of a combination of high- and low-pressure columns is also practiced in the chemical industry for many reasons. One of the benefits of such a column configuration is the ability to reduce the heating duty of the overall column configuration by using the condenser stream of the high-pressure column to supply the heating requirement of the reboiler in the low-pressure column, as illustrated in Figure 11.6. This type of heat integrated system is also used in cryogenic separation processes. The objective of the heat integration is to use the free "cooling" power of the low-pressure column to drive the condenser of the high-pressure column, which otherwise would require a cooling medium that is energy-intensive to create.

In addition, the distillate from the high-pressure column can also be flashed into the low-pressure column, which further results in heat integration. At the same time, such significant heat integration would bring economic benefits in terms of process operations and control this type of heat and mass integration results in a loss of control and degrees of freedom. The condenser pressure of the high-pressure column is now coupled to the bottom temperature of the low-pressure columns. This is because any fluctuations in the low-pressure column's bottoms temperature will affect the cooling duty of the high-pressure column's condenser. In a total condensing configuration (Chapter 4), this would directly impact the amount of condensation, which in turn influences the pressure at the high-pressure column. To this end, the condenser pressure at the high-pressure column is no longer directly controlled. Instead, it is linked to the condenser pressure of the high-pressure column.

Similarly, the temperature at the bottom of the low-pressure column is linked to the condenser's temperature (and based on VLE, the pressure)

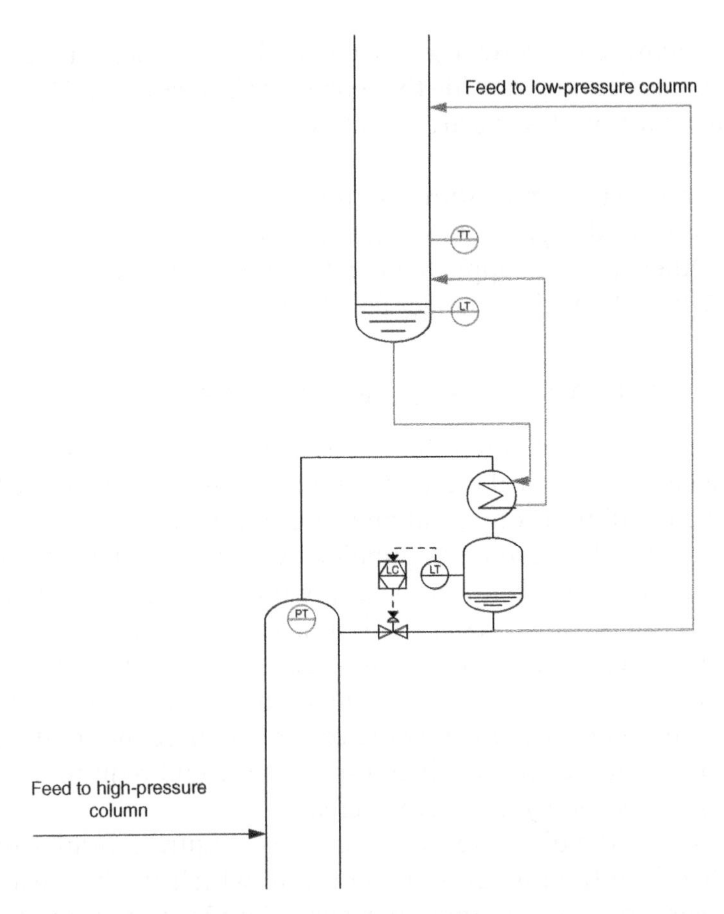

Figure 11.6 High-pressure/low-pressure columns.

of the high-pressure column. This change also influences the level in the low-pressure column bottoms. In principle, it can be controlled by manipulating the bottoms flow rate as illustrated in Figure 11.6. However, the level of integration means the turning characteristics of this loop might need to be adopted to aid the overall cause.

Overall, in this configuration, the otherwise independent condenser pressure of the high-pressure column and the bottoms temperature of the low-pressure column are now interlinked and cannot be controlled independently. As a result, variations and disturbances will propagate through the twin column configuration. This complexity can be dealt with by considering the overall configuration as a single column. As a result, controlling the low-pressure column condenser and high-pressure column reboiler ensures the overall mass and energy balance.

11.2.4 Mechanical Vapor Recompression

In comparison to the examples listed above, this configuration is seldom used in distillation operations. In this situation, the condenser vapor is sent through a compressor to raise the temperature of the condenser vapor allowing this “heated” vapor to now be condensed in the reboiler (such as steam would be condensed), providing sufficient reboiler duty. The condenser stream can then be expanded to the previous pressure and condensed again before entering the reflux drum. The overall concept is illustrated in Figure 11.7. The reason for employing this complex configuration is to recover the heat of evaporation of the condenser vapor stream, which would otherwise be cooled down with water.

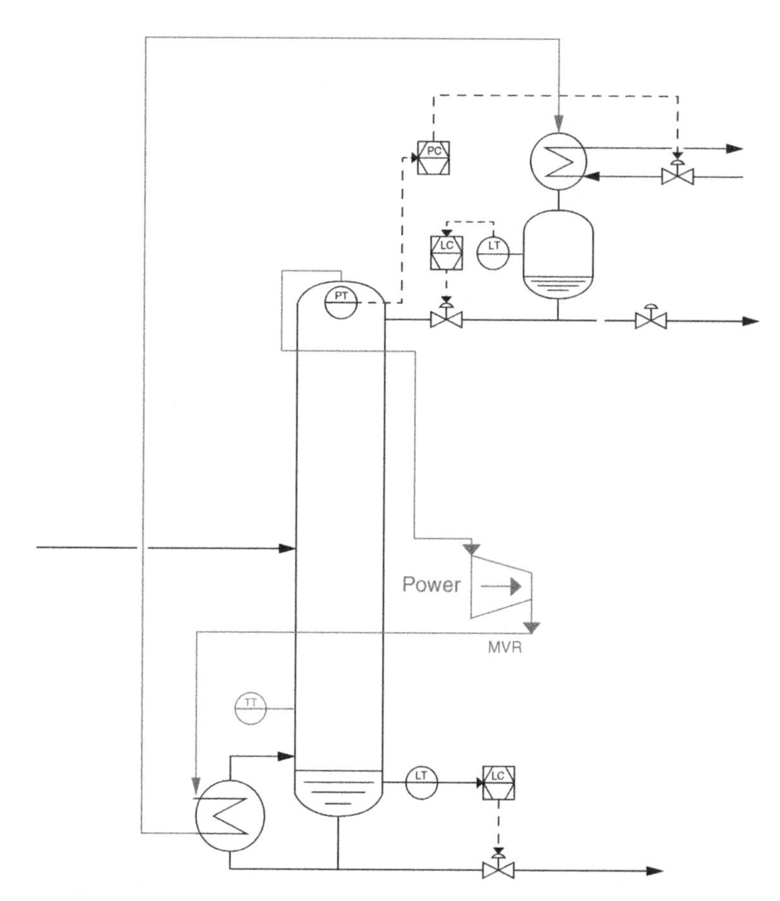

Figure 11.7 Mechanical vapor recompression.

However, the introduction of this heat integration concept adds a level of complexity in terms of operations. In particular, the bottoms reboiler duty is linked to the condenser flow rate and to some degree the temperature (hence the pressure) at the top of the column. In theory, the power to the MVR compressor can be varied as a manipulated variable to control the bottom temperature in the columns. But in practice, quick variations to the compressor power might be limited by the design limitations of such a compressor.

Overall, these examples illustrate the loss of degrees of freedom and that disturbances propagate due to heat integration practiced. As a result, the control of these types of integrated column designs is much more complicated than controlling stand-alone distillation systems.

In general, the following tools and considerations can be useful to make an informed judgment of a given situation case:

- Clearly understand the operational priorities at a plant-wide level and breaking these operational priorities (including economics) down to priorities and requirements that must be met a column level.
- Understanding the column design and its fundamental operations.
- Reassessing of the MV/CV pairings through tools such as relative gain analysis.
- The alteration of both control structure and tuning at a regulatory and advanced regulatory control level in accordance to the operational priorities and design limitations.
- Employing advanced regulatory and model-predictive control layers to enable on-specification and robust overall control.
- If on-specification and robust control is deemed infeasible, then investigate other operational changes such as increasing product variability specifications, lowering set points and the ability to blend/buffer products.

11.3 Materials Recycling

Distillation columns can also be used in a plant-wide context to separate raw materials from products such that the unreacted raw materials can be recycled back to a reactor, as illustrated in Figure 11.8. Such a process design aims to improve the conversion efficiency.

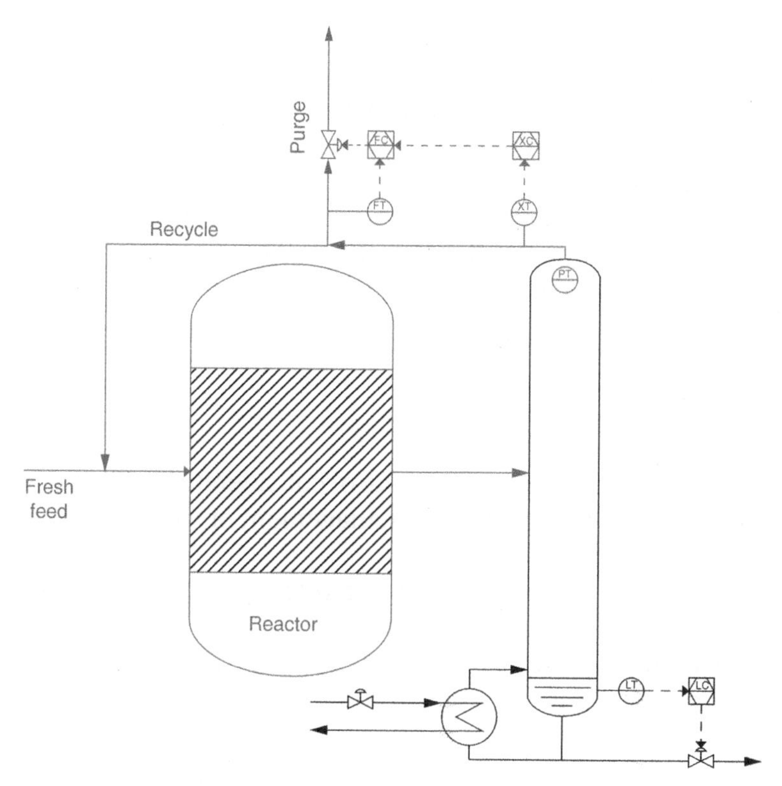

Figure 11.8 Materials recycle loop.

A similar design principle can also be used to recycle undesired products formed during a reaction, thereby reducing the waste produced during production processes. The major challenge presented in this type of endeavor is the overall "snow-ball" effect. That is any minor variations in the recycle feed stream can lead to an incremental accumulation or depletion, which then results in more significant issues. One way to tackle this type of shortcoming is to explicitly introduce a purge stream where the incremental accumulation or depletions can be rejected out of the recycle loop. As illustrated in Figure 11.8, this is achieved by introducing a dedicated cascade loop where the purge flow rate is controlled and is periodically updated based on the composition of key components accumulated in the recycle stream. Typically, components such as inserts, as well as trace side products, are controlled through the design of such a purge system. In comparison, a purge system can also be used to control an overall mass balance and control the conversion rate of all feeds. That is a smaller purge flow will inevitably lead to increased conversion efficiency and increased levels of inert and other trace compounds trapped in the recycle loop.

The system pressure in the recycle loop will set the pressure at the condenser in this type of situation. In turn, the system pressure will be affected by the buildup of inerts and conversion rates. The objective is to obtain overall improved conversion efficiency and quality. The challenge is to avoid the accumulation of components leading to the snow-ball effect.

Tutorial and Self-Study Questions

11.1 Summarize key plant-wide control considerations.

11.2 Describe the movement and management of variation for plant-wide control.

11.3 Describe short-term and long-term plant wide variation management and control strategies.

11.4 Articulate the cascaded unit's plant-wide control rules.

11.5 Describe the "snowball effect" and how it occurs.

11.6 Articulate Luyben's plant-wide control rules.

11.7 What has occurred RE: plant design practices in the past 20+ years that makes process control more vital (i.e. requiring more focused attention earlier in project design cycles) than ever before?

11.8 Should all process control and automation (not *JUST* the Safety Instrumented Systems [SIS]) be considered part of your company's process safety management (PSM) activities? Why? Is it?

11.9 Figure 11.9 shows a simplified process flow diagram for the production of benzene via the hydrodealkylation of toluene. The fresh and recycled toluene is mixed with hydrogen first. The mixture is heated up in the furnace (F-100) at a sufficiently high temperature to initiate the reaction. The mixture is then fed to the reactor (R-100). The effluent, quenched by the cooler (E-100), is then sent to the separator (V-100). The gas phase consists of a hydrogen/methane mixture with small amounts of benzene and toluene. The liquid phase collects benzene and toluene, as well as heavy components and dissolved lights. The gas is recycled to the reactor via the compressor (C-100) and afterward mixed with fresh hydrogen. The liquid is sent to the distillation column

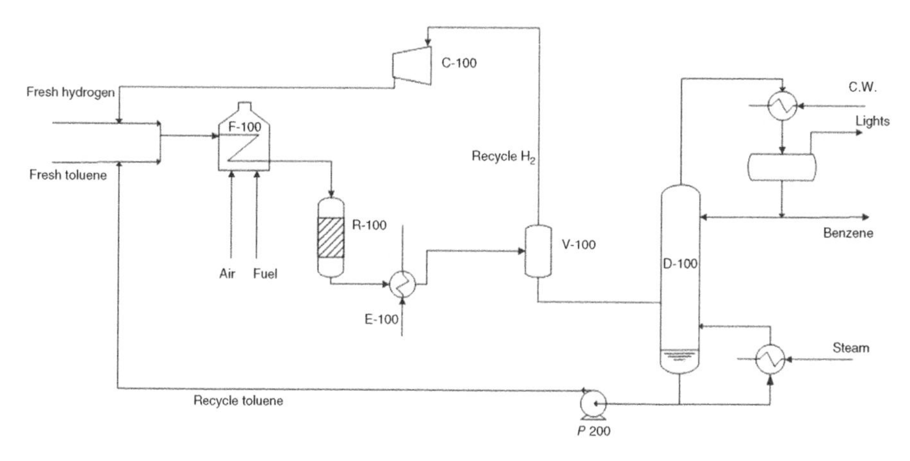

Figure 11.9 A simplified process flow diagram for the production of benzene via the hydrodealkylation of toluene.

(D-100), in which the lights, product (Benzene), and unreacted toluene are separated. The unreacted toluene is recycled back to the fresh toluene stream.

Your task is to design a control scheme for this process. Determine the control degrees of freedom (CDoF). Determine the controlled variables (CVs) and define one suitable manipulated variable (MV) for each CV (use all CDoF).

- What quality variables do you consider in your control design?
- What is your inventory control plan (i.e. columns and separator levels)?
- How would you set the production rate for this plant?

Re-draw the process flow diagram of Figure 11.9 in your answer. Draw your proposed control scheme using standard symbols (including controllers and control valves).

Is there a snowball effect in your design? Explain your answer.

References

Buckley, P.S. (1964). *Techniques of Process Control*. New York: Wiley.

Luyben, W.L. (1993a). Dynamics and control of recycle streams. 1. Simple open-loop and closed-loop systems. *Industrial & Engineering Chemistry Research* 32: 466–475.

Luyben, W.L. (1993b). Dynamics and control of recycle streams. 2. Comparison of alternative process designs. *Industrial & Engineering Chemistry Research* 32: 476–486.

Luyben, W.L. (1993c). Dynamics and control of recycle streams. 3. Alternative process designs in a ternary system. *Industrial & Engineering Chemistry Research* 32: 1142–1153.

Luyben, M.L., Tyreus, B.D., and Luyben, W.L. (1997). Plant wide control design procedure. *AIChE Journal* 43 (12): 3161–3174.

Luyben, W.L., Tyreus, B.D., and Luyben, M.L. (1998). *Plant Wide Process Control*. New York: McGraw-Hill.

Shinskey, F.G. (1996). *Process Control Systems: Application, Design, and Tuning*, 4e. New York: McGraw-Hill.

Shinskey, F.G. (1997). Averaging level control. *Chemical Processing* 60 (9): 58.

Svrcek, W.Y., Mahoney, D.P., and Young, B.R. (2014). *A Real-Time Approach to Process Control*, 4e. Chichester, UK: Wiley.

Tyreus, B.D. and Luyben, W.L. (1993). Dynamics and control of recycle streams. 4. Ternary systems with one or two recycle streams. *Industrial & Engineering Chemistry Research* 32: 1154–1162.

Workshop 1 Hands-on Learning By Doing

Course Philosophy: *"Learning By Doing"* or *"Hands-on Learning"*

***In conjunction with this workshop, you should review the Preface and Chapter** 1.*

This section of the book consists of a set of learning modules or workshops, each of which is intended to enhance your understanding of distillation process control and simulation theory and application through hands-on experience using the latest simulation technology for actual virtual laboratory experiments. For these workshops, we recommend using a commercial dynamic process simulator such as Aspen HYSYS or Schlumberger's Symmetry, as indeed several of the workshop exercises are built on case studies that are bundled with these process simulators. We will indeed in the following instructions as appropriate.

Key Learning Objectives

1. Develop a working knowledge of the process simulator steady-state and dynamic simulation package.
2. Understand the fundamentals of steady-state and dynamic process simulation.

A Real-time Approach to Distillation Process Control, First edition. Brent R. Young, Michael A. Taube, and Isuru A. Udugama.

Companion website: www.wiley.com/go/Young/DistillationProcessControl

Book Coverage

This set (or course) of workshop deals with the fundamental and underlying principles of automatic distillation process control and simulation. The book covers the theory associated with single input/single output (SISO) loops and how these are configured into multi-loop schemes to control complex distillation columns and entire plants including distillation. It will not discuss in detail the hardware or individual measurement techniques, such as flow, temperature, pressure, and so on, except as the measurement affects the control loop. However, Chapter 3 does provide a simplified summary of control loop hardware.

Prerequisites

No elaborate prerequisites are required for studying the topic of distillation process control using a real-time approach; however, an understanding of unit operation modeling is assumed. It is inevitable that modeling involving differential equations will, by necessity, be involved in parts of the theory and workshops. Quite often mathematics is a barrier that prevents a clear understanding of control concepts and implementation of process control theory. The real-time approach will remove or, at least, minimize these barriers.

Study Material

This book and a suitable dynamic process simulator (including user manual) are the only materials required for the topic. The book chapters and these workshops are independent, unique, stand-alone, and specifically designed for the tutorial/workshop approach.

The references that are provided at the end of each chapter detail additional selected study and reference reading. The available literature on the subject of process dynamics and control is massive. Additional literature is readily available from instrumentation vendors such as Emerson, Honeywell, and Rockwell. This vibrant area of chemical engineering is represented by the Instrumentation, Systems, and Automation Society (ISA).

Organization

The implied course of study in this book consists of workshops associated with each of the chapters. Each workshop has associated with it a specific portion of the book that provides the necessary theoretical background. During the workshop, a specified assignment using the dynamic process simulator is to be completed.

The Simulation Tool

There is a general procedure in building cases in commercial dynamic process simulators such as Aspen HYSYS and Schlumberger's Symmetry. First of all, you need to construct the basic unit operations in steady state, such as streams, valves, separators, columns/towers, heaters, and coolers. You have to establish a proper pressure difference along the process pathway and obey pressure–flow relationship rules. You also need to specify the opening of the valve and note down the Cv values for the valve. You should not include dynamic factors into the system such as inventory capacitance, controllers, and selector blocks. Once you have solved your steady-state case, save your case for the steady-state model and you can then convert to dynamics.

The general rules for converting a model from steady state to dynamics are as follows:

1. Enter dynamic mode.
2. Size all relevant process equipment.
3. Add the valves and controllers to control the simulation.
   i. Valves must be sized, and controllers must be properly set up.
   ii. Use selector blocks to simulate disturbances, if desired.
4. Add the strip charts to monitor the process.
5. Set up the pressure/flow specification on all boundary streams.
6. Run the dynamics assistant.
7. Save your case (it is a good idea to save your case quite often when working in dynamics as the simulator is not able to go backward in simulation time; the only way to go back is by using a saved case).
8. Start the integrator.

Note that these rules apply to all cases. You must have your steady-state case solved prior to converting to dynamics. Otherwise the system might be unstable.

Overall Learning Objectives

1. Understand the basic theoretical concepts of feedback and SISO loops.
2. Understand the components of a control loop and how they interact.
3. Understand process control terminology.
4. Have an appreciation of process dynamics.
5. Know how to develop first-order plus dead time models for processes described by these dynamics.
6. Know how to tune controllers.
7. Know how and where to implement such techniques as cascade, feedforward, ratio, dead time, and multi-loop control.
8. Appreciate the use of process simulation in the development and validation of control strategies.
9. Develop an understanding of the distillation unit operation control schemes.
10. Understand what is meant by plant-wide control and be able to implement a plant-wide control strategy including distillation columns.
11. Familiarize yourself with the appropriate simulation software.

Tasks

1 – Get Familiar with the Simulator

You are recommended to read over the "Getting Started with Aspen HYSYS" or "Symmetry Fundamentals" chapter in the appropriate simulation user manual if you are new to chemical process simulation software. This will provide you with the basic ideas of simulation, unit operations, and degrees of freedom. You are not required to follow all the tutorials in the user manual.

2 – Steady-State Tutorial

In this task you are instructed to build your first steady-state model. Refer to the HYSYS "Sweet Gas Refrigeration Plant" or Symmetry "Dew Point Gas Plant" tutorials from the appropriate user manual and follow the steps as instructed. Gain the idea of how to set up a simulation file, how to create new unit operations and material flow streams, and how to define them. Understand the key points of solving a steady-state simulation case.

Once you have finished, refer to the example file in the simulator case study documentation folder. Is the data in each unit operation and stream the same?

3 – Transitioning from Steady State to Dynamics Tutorial

In this task, you are instructed to convert a solved steady-state simulation model into dynamics, which is essential in your future workshops. Follow the tutorial instructions on Aspen "Dynamic Simulation" or "Process Control Update" in Symmetry from the respective manual. Learn how to plot variables with respect to time in the simulator as well as how to set up controllers in dynamics.

If you are interested, read the Aspen "Pressure Flow Specifications" section or continue with the Symmetry "Dynamic Specification Analysis" tutorial. This would illustrate how dynamic specification differs from steady-state specification and how to solve specification problems in dynamics.

Tutorial and Self-Study Questions

1.1 Describe what is meant by the term degrees of freedom in simulation.

1.2 Why are pressure-flow relationships important in dynamic simulations?

1.3 How are the dynamic pressure-flow relationships are defined for unit operations in process simulators such as HYSYS or Symmetry?

1.4 What is the general procedure for constructing dynamic simulations in process simulators such as HYSYS or Symmetry?

Workshop 2 Fundamental Distillation Column Control

Introduction

Prior to attempting this workshop, you should review Chapters 1–5 and have completed Workshop 1. You should understand how to start, stop, and make process changes to a working dynamic model and how to add control loops to a dynamic model.

In this workshop, you will:

1. Convert a steady-state distillation column model into a dynamic column model
2. Add several control strategies to the dynamic model
3. Test the control strategies for several different load and set-point changes

Key Learning Objectives

This workshop will demonstrate the impact the selection of a control strategy can have on the operability and controllability of a simple distillation column. In addition, this workshop will reinforce your knowledge of converting steady-state simulations into dynamic ones. You will also use a dynamic column operation "in anger" for the first time.

A Real-time Approach to Distillation Process Control, First edition. Brent R. Young, Michael A. Taube, and Isuru A. Udugama.

Companion website: www.wiley.com/go/Young/DistillationProcessControl

Tasks:

1- ***Steady State:***

The distillation column built here will be a 15-stage debutanizer column. There are two feed streams, each of which is a mixture of light hydrocarbons (C3 through C8). The top product will be a butane-rich stream, and the bottom product will be a mixture of hydrocarbons heavier than butane.

A process overview is given in Figure W2.1.

Build a simulation model of this process with the following specifications:

In the basis environment, add a fluid package with:

- Property Package: Peng-Robinson
- Components: Propane (C3), *i*-butane (*i*C4), *n*-butane (*n*C4), *i*-butene (*i*C4=), *i*-pentane (*i*C5), *n*-pentane (*n*C5), *n*-hexane (*n*C6), *n*-heptane (*n*C7), *n*-octane (*n*C8).

Switch to the simulation environment and add two feed streams with the pressure, temperature, and flow data that are shown in Figure 2.1 and the following mass fraction data:

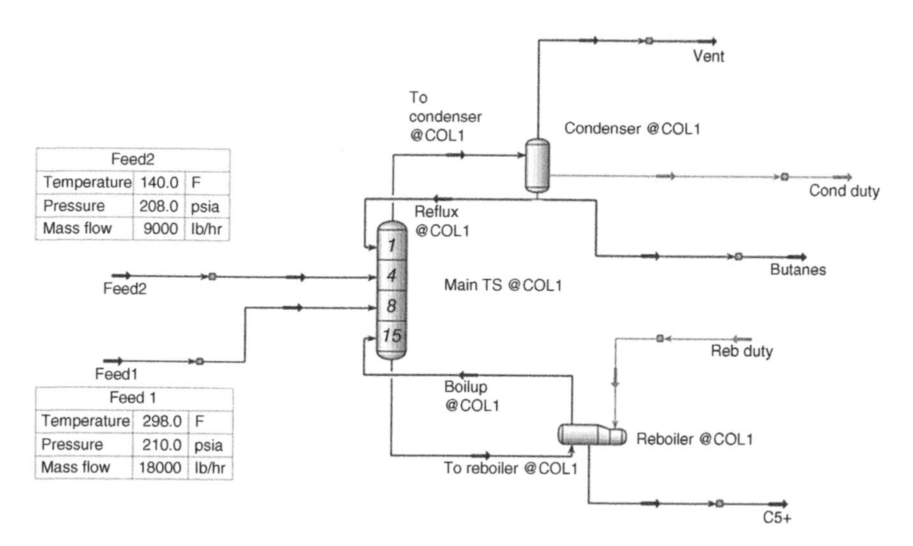

Figure W2.1 Process overview.

- Feed1: 0.012 C3, 0.17 *i*C4, 0.17 *n*C4, 0.008 *i*C4=, 0.14 *i*C5, 0.14 *n*C5, 0.11 *n*C6, 0.13 *n*C7, 0.12 *n*C8
- Feed2: 0.02 C3, 0.19 *i*C4, 0.2 *n*C4, 0.22 *i*C4=, 0.16 *i*C5, 0.21 *n*C5, zero *n*C6, zero *n*C7, zero *n*C8

Add a distillation column to your flow sheet with the configuration as indicated in Figure 2.1. Use the following steady-state specifications for the column:

- Partial condenser, zero overhead vapor rate, condenser pressure: 1413 kPa, zero condenser and reboiler pressure drops, 1448 kPa reboiler pressure, 0.95 butane recovery, and 0.05 mole fraction C5s in the distillate.
- Good estimates for initial convergence are: 2.299 reflux ratio, 94.67 kg-mol/h distillate rate, 87 °C condenser temperature, and 160 °C reboiler temperature.

Now the model should be totally solved in steady state.

Explore the results from the converged steady-state solution by answering the following questions:

- Confirm what are the current specifications for the column?
- What is the reflux ratio of the column with the current specifications?

Change the butane recovery fraction to 0.9625 and the amount of C5s in Top to 0.025.

- What is the reflux ratio now?
- What is the molar reflux flow?

Make this an active specification with the current value as the spec value and deactivate the C5s in Top specification.

Now test the response of the column to changes in the flows of the feed streams. Record the tray 6 temperature, mass fraction of *i*C5 in the butanes stream, condenser duty, and reboiler duty for the following (Feed1, Feed2) flow combinations in pounds per hour: (18 000, 9000), (9000, 18 000), (0, 27 000), (27 000, 0).

When you are finished, return the flows for the two feed streams to their original values. You are now ready to begin to transform the steady-state case into a dynamic model.

2- ***Converting to Dynamics:***

In Workshop 1, we converted a steady-state system to a dynamic model. We are going to do the same thing here. Converting a distillation column is a little more complicated, but the instructor should be prepared to help you get through any difficulties that arise.

Just as with Workshop 1, there are several basic steps that are required to convert from steady state to dynamics:

- Adding the valves on the feed and product streams. Note that we will not add a valve on the vent stream. Instead, we will use a flow spec instead of the standard pressure spec.
- Define the P/F specs for all feed and product streams. We will also set the mass flow of reflux as a dynamic spec. This is a column requirement.
- Set various equipment volumes.
- Set up the various PID controllers. We will use six controllers in this simulation.

All of these steps must be done before the case can be run dynamically. In addition, a strip chart will be set up to monitor the process.

We want to add four valves to our simulation. All valves will have a steady-state pressure drop of 7 psi:

- VLV-100: Inlet = To Feed1; Outlet = Feed1
- VLV-101: Inlet = To Feed2; Outlet = Feed2
- VLV-102: Inlet = Butanes; Outlet = Butane Product
- VLV-103: Inlet = C5+; Outlet = Liquid Product

The valves on the Feed1 and Feed2 streams are not automatically sized by the simulator. Use the process simulator to manually calculate and enter the Cv values.

For the following streams, activate the pressure specification. Use the steady-state pressure in all cases.

- To Feed1
- To Feed2

- Butane product
- Liquid product

For the following streams, activate the flow specification. Again, use the steady-state flows.

- Vent
- Reflux

In order to set the flow spec on the reflux stream, you will have to enter the column environment.

The boundary streams of our simulation changed when we added the valves to the simulation. For this reason, you may have to deactivate the dynamic specifications for the Feed1 and Feed2 streams.

There are three pieces of equipment that require sizing information before running dynamically: the condenser, reboiler, and tray section.

We will set the volume of both the condenser and the reboiler to 530 ft^3.

For the tray section, specify the following values:

Tray diameter	4.5 ft
Weir height	0.15 ft
Weir length	4.0 ft
Tray spacing	1.8 ft

We need to have six controllers in our simulation. The following variables will be controlled using PID controllers:

- Feed1 – Mass flow
- Feed2 – Mass flow
- Condenser – Liquid percent level
- Condenser – Vessel pressure
- Reboiler – Liquid percent level
- Main TS – Stage 6 temperature

The mass flow rates of the two feed streams are easily controlled with the valves that are on the streams. Use appropriate controller tuning parameters for flow.

Control the liquid level in the condenser by adjusting the valve on the butanes stream. Control the liquid level in the reboiler by adjusting the valve on the C5+ stream.

Manipulating the condenser duty heat flow will control the pressure in the condenser, which sets the pressure in the rest of the column. Increasing the cooling duty will cause more condensation inside the condenser, hence the pressure will drop. Conversely, decreasing the cooling duty will lead to an increase in the column's pressure. In order to define the operation of this valve, you will need to specify the controller's range, and the range of the duty valve for the condenser. Size the condenser's duty valve using the direct Q method. Enter a maximum heating value that is approximately twice the steady-state value.

The temperature of the sixth tray in the column will be controlled by manipulating the reboiler duty. This tray was chosen because it showed a good response to a change in the reboiler duty. Again, size the duty valve using the direct Q method and use a maximum value that is twice the steady-state value. You will need to set the range of the Temperature Controller (TIC) also.

The actions that the controllers require have not been given here. At this point in the course, you should be able to determine the correct controller action yourself. However, if you have any questions, do not hesitate to ask the instructor for help.

A process simulator P&ID diagram is included as Figure W2.2 to assist you.

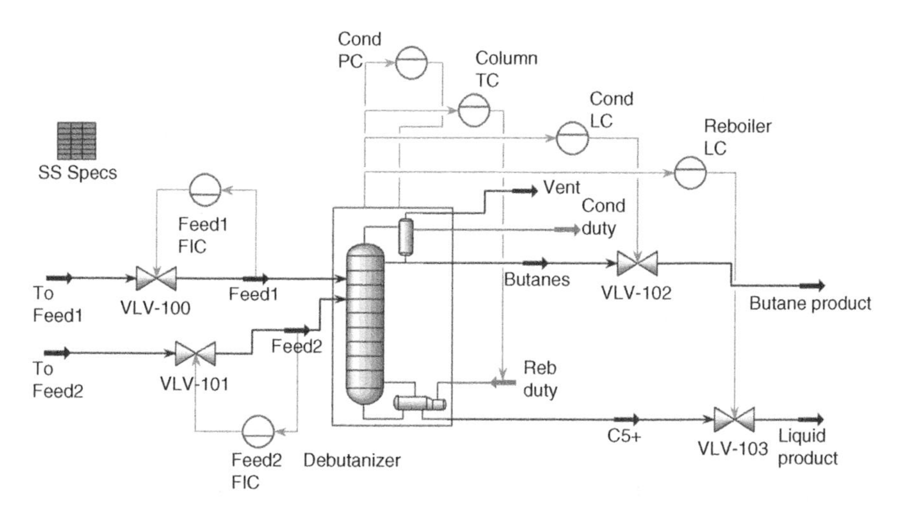

Figure W2.2 Simulator P&ID diagram (LV-1 control strategy, indirect feed).

Strip charts need to be included in the case for your use to dynamically monitor key variables. Set up a compositions strip chart to track the total mass fraction of C4s (i.e. sum of the mass fractions for *i*-butane, *n*-butane, and *i*-butene) and C5s (i.e. *i*-pentane and *n*-pentane) in the butanes stream. The mass fractions may be summed using a spreadsheet unit operation in the process simulator.

At this stage, it is very important to save your case!

As was done in Workshop 1, the Dynamics Assistant will be used to check the simulation over and to ensure that all of the required dynamic specifications are set.

Remember that the Assistant will only test your simulation against default rules. For this reason, it will suggest that you remove the flow spec on the vent stream. However, we know that this is a good specification in this case, so we can ignore this suggestion.

The assistant will also alert you to the fact that the pressure profile for the column in steady state may not match the profile of the column in dynamics. This, too, is acceptable.

If you followed the steps in this module carefully, the assistant should not give any serious warning messages. If you have any questions about the meaning of any of the messages that the Dynamics Assistant gives, ask the instructor for more information.

We can now enter the dynamic mode and try to run the simulation dynamically.

3- ***Dynamics Exercises:***

Once all of the equipment has been entered in steady state, the P/F specs have been set, and the equipment has been sized, we can enter the dynamic environment in the process simulator. Now, we can start to run the simulation. If everything is set up correctly, the integrator should run without any trouble.

- What is the value of C5s in butanes when the simulation reaches steady state?
- How does this compare to the steady-state solution found earlier?

Make sure that the stage 6 temperature is the same for both cases.

At this stage, it is again very important to save your case, perhaps with a different name!

We have created a dynamic debutanizer model and implemented the LV-1 control strategy. This strategy is often described as an energy balance configuration or the indirect feed-split strategy. In this strategy, the product flows (and hence the material balance of the column) are controlled by level controllers on the reboiler and condenser drum.

To test your controller strategy, perform a variety of load and set-point disturbances on the model. Suggestions for disturbances include:

- Changes in feedflow rates as given in the chart (see above). Note – you may not be able to reach the upper limit on the flow of Feed2 due to the size of the valve. Increase the Cv of the valve to achieve the desired flow rate.
- Changes in feedflow rates also change the overall feed composition into the column unless the two feeds are simultaneously changed by the same ratio. Study the impact of feed composition changes only by changing the feed composition only on one or more of the feeds.
- Change the temperature and/or pressure set points.

Save this case with the LV-1 control strategy.

As discussed in the book chapters, many other possible strategies can be successfully implemented, depending on the column characteristics and control objectives. Now, implement one other control strategy from the chart in Table W2.1 by modifying your existing case. Test your new configuration with the same disturbances listed above and compare and contrast your results.

Control Configuration Notation:

L = Liquid flow down the column, reflux rate
V = Vapor flow up the column, boil up, or reboiler duty
D = Distillate flow rate
B = Bottoms flow rate

If this workshop is part of a course, you will likely be required to present your findings in a short report. Also include in the submission a copy of the simulation files which you used to generate your findings.

Table W2.1 Control configurations for Workshop 2.

Control configuration	mv for condenser level control	mv for reboiler level control	mv for primary composition control (temp on Stage 6)	mv for secondary composition control (fixed in base case)	mv for pressure control
LV-1	Distillate flow rate	Bottoms flow rate	Reboiler duty	Reflux flow rate	Condenser duty
LV-2	Distillate flow rate	Bottoms flow rate	Reflux flow rate	Reboiler duty	Condenser duty
DV-1	Reflux flow rate	Bottoms flow rate	Reboiler duty	Distillate flow rate	Condenser duty
DV-2	Reflux flow rate	Bottoms flow rate	Distillate flow rate	Reboiler duty	Condenser duty
LB-1	Distillate flow rate	Reboiler duty	Bottoms flow rate	Reflux flow rate	Condenser duty
LB-2	Distillate flow rate	Reboiler duty	Reflux flow rate	Bottoms flow rate	Condenser duty

Workshop 3 Distillation Column Model Predictive Control

Introduction

Prior to attempting this workshop, you should review Chapters 5–9 and have completed Workshops 1 and 2. You should understand how to start/stop and make process changes to a working dynamic process model.

This workshop will illustrate how to implement model-predictive control (MPC) on a debutanizer distillation column.

This workshop will use the same column model developed previously in the prior workshop. We will devise an MPC scheme to control both the total C4s in the bottoms stream as well as the total mole fraction of C5s in the distillate stream. Step tests will be performed, and the resulting data used to form the basis of the control strategy.

Key Learning Objectives

This workshop will demonstrate how to perform step tests on a dynamic model. The results are then used to set up an MPC controller. The system is assumed to be a first order model. The user will take the results of the step tests and determine the model parameters – the gain, time constant, and delay.

Description

A debutanizer distillation column was modeled in Workshop 2 (Figure W3.1). At that time, we explored various control strategies that used multiple SISO controllers to control its operation.

A Real-time Approach to Distillation Process Control, First edition. Brent R. Young, Michael A. Taube, and Isuru A. Udugama.

Companion website: www.wiley.com/go/Young/DistillationProcessControl

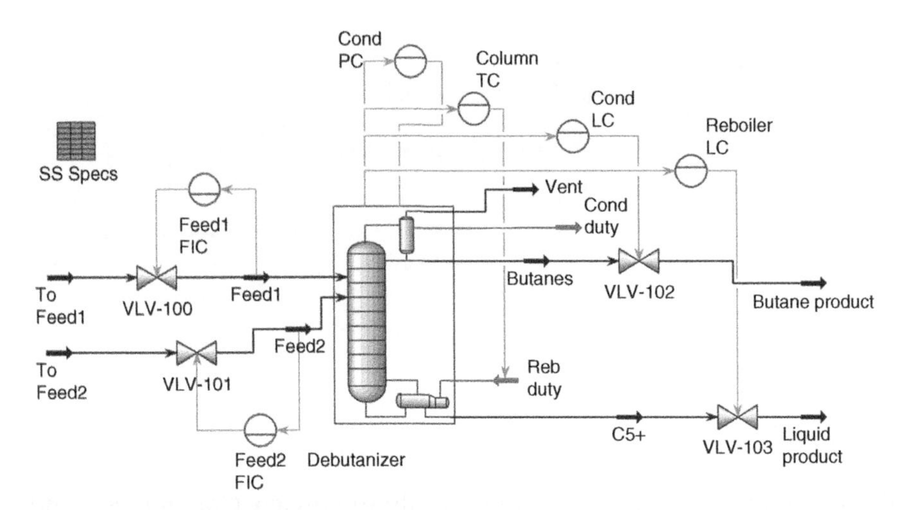

Figure W3.1 Simulator P&ID diagram (LV-1 control strategy, indirect feed).

Now, we will explore the use of a 2×2 MPC controller to control the product compositions in both the distillate and bottoms streams. The product compositions to be controlled will be composite mass fraction variables:

- Total mass fraction of C4s in the bottoms stream
- Total mass fraction of C5s in the distillate stream

The manipulated variables are the reflux flow rate and the reboiler duty.

Tasks

To begin the workshop, open the saved case from the Workshop 2 of a complete dynamic model of the debutanizer with the LV-1 control strategy implemented. A process flowsheet of the model with this controller strategy is shown in Figure W3.1.

For the MPC controller, we want to control the:

- Total mass fraction of C4s in the bottoms stream
- Total mass fraction of C5s in the distillate stream

Both can be calculated in a spreadsheet model in the simulator.

The manipulated variables are the reflux flow rate and the reboiler duty. The reflux flow rate is available from the Debutanizer PFD.

In the simulation, the reboiler duty is controlled by the column TC controller. In preparation for the step testing to design the MPC controller, we must disable this controller.

Why must this be done? How can you do this in your model?

Determine appropriate values for the Control Interval and the Step Response Length. The latter is the number of sampling intervals necessary to reach steady state. The MPC controller allows this range to be 15–100. Dividing the time that the model takes to reach steady state, by the Step Response Length, will give an appropriate Control Interval.

Note – Suggestions for step changes in the manipulated variables are:

Reflux flow rate: −10% decrease
Reboiler duty: −1% decrease

When we set up the MPC, we will use the following notations:

PV1 = C4s in bottoms stream (calculated in a spreadsheet cell)
PV2 = C5s in distillate stream (calculated in a spreadsheet cell)
MV1 = Reflux flow rate
MV2 = Reboiler duty

Make the step changes and record the following data:

For a change in MV1 by −10%
Value of PV1 at $t = 0$ _______ C4 mass fraction
Value of PV1 at $t = \infty$ _______ C4 mass fraction
Value of PV1 at 63.2% of change = _______

a. Value of t at 0.632 PV1 = _______ seconds
b. Dead Time (Delay) (if any) = _______ seconds

Value of PV2 at $t = 0$ _______ C5 mass fraction
Value of PV2 at $t = \infty$ _______ C5 mass fraction
Value of PV2 at 63.2 % of change = _______

a. Value of t at 0.632 PV2 = _______ seconds
b. Dead Time (Delay) (if any) = _______ seconds

For a change in MV2 by −1%

Value of PV1 at $t = 0$ _______ C4 mass fraction
Value of PV1 at $t = \infty$ _______ C4 mass fraction

Value of PV1 at 63.2% of change = _______

a. Value of t at 0.632 PV1 = _______ seconds

b. Dead Time (Delay) (if any) = _______ seconds

Value of PV2 at $t = 0$ _______ C5 mass fraction

Value of PV2 at $t = \infty$ _______ C5 mass fraction

Value of PV2 at 63.2% of change = _______

a. Value of t at 0.632 PV2 = _______ seconds

b. Dead Time (Delay) (if any) = _______ seconds

What are the controllers' ranges?

PV1Min = _______

PV1Max = _______

PV2Min = _______

PV2Max = _______

From the data collected, the process model variables are calculated. The gain must be made dimensionless. The gain, K_p, is defined as follows:

$$K_p = \left\{ \frac{(PV_f - PV_i)/(PV_{max} - PV_{min}) * 100\%}{(MV_f - MV_i)/(MV_{max} - MV_{min}) * 100\%} \right\}$$

In our case, the denominator is equal to the change in the valve output (−10%).

It is now suggested that you summarize the model parameters in a Table such as Table W3.1

Table W3.1 Debutanizer MPC model parameters.

Gi.j refers to change in PV_i due to a change in MV_j			
	K_p (Gain)	T_p (time constant)	T_d (time delay)
G1.1			
G1.2			
G2.1			
G2.2			

We now have enough information to create the MPC. Return to the simulator case and add an MPC controller to the simulation. Set up the following parameters: number of inputs (2), number of outputs (2), the control interval (determined earlier), the step response length (100), and select first-order model. Remember to use the same ranges that were used to normalize the data. For consistency with our earlier calculations, recall that we are using the following notation:

PV1 = C4s in bottoms stream (calculated in a spreadsheet cell)
PV2 = C5s in distillate stream (calculated in a spreadsheet cell)
OP1 = Reflux flow rate
OP2 = Reboiler duty

The MPC controller setup is now complete. Place the controller into Auto.

At this point, the user is free to try out the MPC. Make changes to the feed flow and the controller set points. Observe responses. Is the control significantly better than the previous LV-1 control configuration?

If this workshop is part of a course, you will likely be required to present your findings in a short report. Also include in the submitted a copy of the simulation files which you used to generate your findings.

Workshop 4 Distillation Column Control in a Plant-Wide Setting

Introduction

Prior to attempting this workshop, you should review Chapters 7, 10, and 11 and have completed Workshops 1–3. You should understand how to start/stop and make process changes to a working dynamic process model.

In this workshop, you will get a chance to use a plant-wide control design procedure to design a suitable strategy for an isomerization process. Then, you will test the control strategy for several perturbations using a dynamic model of the process.

Key Learning Objectives

In this workshop, you will learn how to use the steps of a plant-wide process control design procedure to design a robust process control strategy for an entire process, not just a single unit operation. In addition, you will reinforce your knowledge of using the process simulator to design and test control strategies.

Description

We will use a simple isomerization process in this workshop. The isomerization process converts normal butane (*n*C4) into isobutane (*i*C4). The process consists of a reactor, two distillation columns, and a liquid recycle stream. Details on each major unit is given below.

A Real-time Approach to Distillation Process Control, First edition. Brent R. Young, Michael A. Taube, and Isuru A. Udugama.

Companion website: www.wiley.com/go/Young/DistillationProcessControl

The reaction carried out in the plug flow reactor is:

$$nC4 \rightarrow iC4 \qquad \text{(W4.1)}$$

This vapor-phase reaction is irreversible and first order with respect to nC4. To achieve good reaction rates, the reactor is run at elevated temperatures (~400 °F) and pressures (~600 psia). This exothermic heat of reaction (−3600 BTU/lb-mol) causes a temperature rise (~30 °F) across the adiabatic reactor. This generated heat must be removed, or the reactor temperature will increase and cause reactor runaway.

The large de-*iso*-butanizer (DIB) column separates the C4s. The iso/normal separation is difficult because of the similar relative volatilities, so the column has many trays (50), a high reflux ratio (~7), and a large diameter (16 ft). The column has two feeds – the reactor effluent stream after it is cooled and condensed and a fresh feed stream consisting of four components (propane, iC4, nC4, iC5). The DIB column operates at 100 psia so that cooling water can be used in the condenser (reflux drum temperature is 124 °F). The reboiler temperature is ~150 °F, so low-pressure steam is used for the reboiler duty.

The distillate product from the DIB column is the isobutane product stream. It has a specification of a maximum of 2 mol% nC4. Also, any propane impurity in the fresh feed stream is removed in the distillate product stream. The bottoms product from the DIB column contains most of the nC4 and all of the heavy iC5 impurity from the fresh feed.

The purge column is used to remove the iC5 impurity. It is a much smaller column, with 20 trays and only 6 ft in diameter. The iC5 impurity and some nC4 are removed in a small bottoms stream (~30 lbmol/h flowrate). The distillate stream from the purge column is the recycle stream feed to the reactor. It is pumped up to the required pressure and heated through a feed/effluent heat exchanger and a furnace before entering the reactor in the vapor phase.

A process overview is given below in Figure W4.1.

Tasks

To develop the control strategy for this isomerization process, we will use the nine steps of the plant-wide process control design procedure as outlined by Luyben and colleagues in their book, *Plantwide Process Control* (1999).

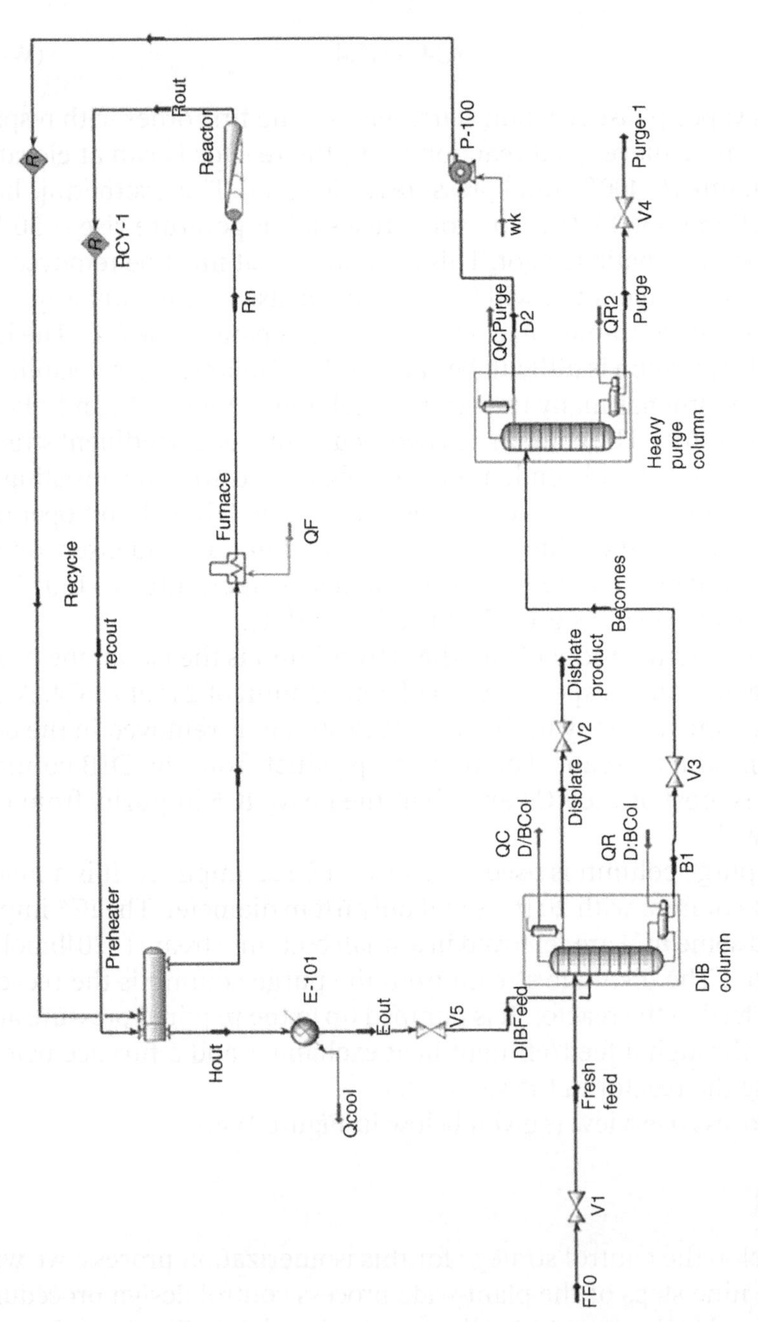

Figure W4.1 Simulator P&ID diagram of the isomerization process.

The nine steps of Luyben's heuristic procedure for plant-wide process control are summarized below:

1. Assess the steady-state design and dynamic control objectives for the process.
2. Count the number of control valves available.
3. Make sure that energy disturbances do not propagate throughout the system by transferring the variability to the plant utility system.
4. Establish the variables that dominate the productivity of the reactor and determine the most appropriate manipulator to control production rate.
5. Select the "best" valves to control each of the product quality, safety, and environmental variables.
6. Fix a flow in every recycle loop and then select the best manipulated variables to control inventories.
7. Identify how chemical components enter, leave, and are generated or consumed in the process.
8. Establish the control loops necessary to operate each of the individual unit operations.
9. Establish the best way to use the remaining control degrees of freedom.

To assist you in this workshop, use Tables W4.1 and W4.2 to list your control strategy decisions. On it, list your control objectives for the process and all available control degrees of freedom (or valves that you can manipulate). Then, using the criteria above, design a control strategy by pairing process variables you want to control with appropriate manipulated variables.

Obviously, there are many possible control strategies for this simple process. However, some will work well, and others will not if they violate the plant-wide design principles.

Table W4.1 Plant-wide control workshop control objectives.

Control objectives
1.
2.
3.
4.

Table W4.2 Pairing of manipulated and controlled variables.

Number	Step	Control loop	Manipulated variables	Controlled variable (or process variable)
1.				
2.				
3.				
4.				
5.				
6.				
7.				
8.				
9.				
10.				
11.				
12.				
13.				
14.				

To study the robustness of the control strategy, test the model using several disturbances. Suggestions to try, include the following:

- Change in reactor inlet temperature set point by 10–20 °F in either direction.
- Change in feed composition (e.g. increased *i*C5 impurity and/or change in ratio of *i*C4 to *n*C4 in the feed).
- Increase in the recycle flow controller set-point by 10–25%.
- Failure of a major utility stream "valve" (cooling water, furnace fuel, steam to a reboiler) by placing the appropriate controller on manual and setting the OP variable to fixed value.

If you have time, you may want to convert the case to model other control strategies by editing the controller configurations, and then test these strategies against the same disturbances above. Be sure to save your edited cases under new names, so you can easily compare the results from several cases.

If this workshop is part of a course, you will likely be required to present your findings in a short report using. Also include on the submitted media a copy of the simulation files which you used to generate your findings.

Reference

Luyben, W.L., Tyreus, B.D., and Luyben, M.L. (1999). *Plantwide Process Control*. McGraw-Hill.

Appendix A

P&ID Symbols

Symbol	Description
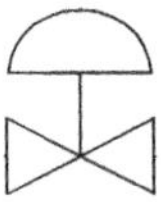	Control valve
	Valve
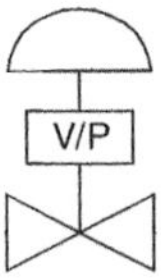 V/P	Control valve with valve positioner
	Check valve
	Pressure relief valve
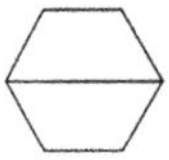	Controller
	Transmitter/sensor

A Real-time Approach to Distillation Process Control, First edition. Brent R. Young, Michael A. Taube, and Isuru A. Udugama.

Companion website: www.wiley.com/go/Young/DistillationProcessControl

Symbol	Description
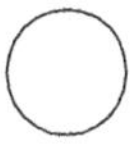	Transmitter/sensor
 Σ	Controller summer
	Summer or multiplier
 ÷	Divider
 <	Selector
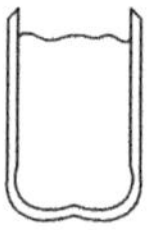	Reactor with cooling jacket
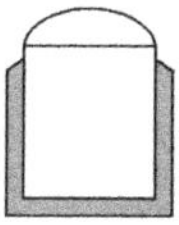	Reactor with cooling jacket
	Knock out drum
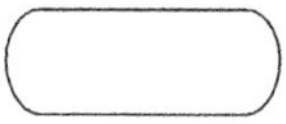	Reflux drum
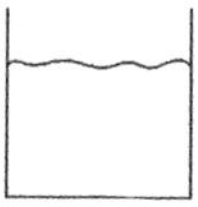	Tank
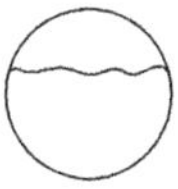	Horizontal tank
	Vertical tank

Symbol	Description
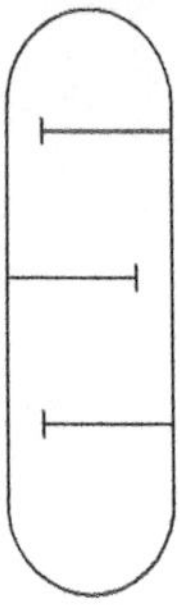	Distillation column tray section
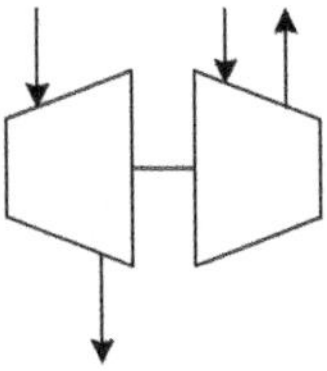	Compressor
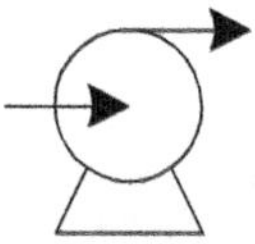	Pump
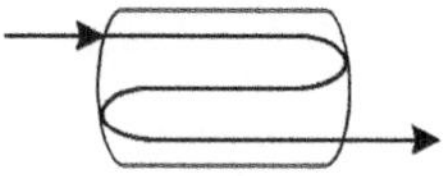	Heat exchanger
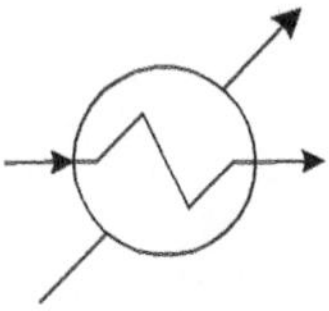	Heat exchanger
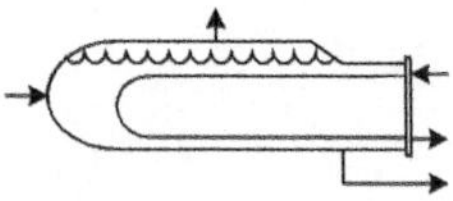	Kettle reboiler
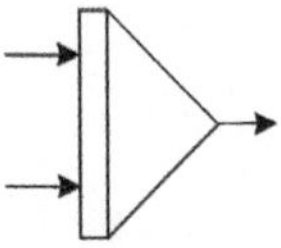	Stream mixer

Index

A Real-time Approach to Distillation Process Control, First edition. Brent R. Young, Michael A. Taube, and Isuru A. Udugama.

Companion website: www.wiley.com/go/Young/DistillationProcessControl